GRAMMAR AND STYLE AT WORK

Second Edition

Don Klepp
Mykon Communications

---

10 9 8 7 6 5 4 3 V0UD

PEARSON

ISBN 978-0-20-596468-0

PREFACE

Grammar and Style at Work responds to a need for supplementary material and exercises to help writers improve their skills. This workbook to that text focuses on grammar, sentence structure, vocabulary and usage, and mechanics such as spelling, punctuation, and numbers. The book provides self-directed instruction and exercises that students can use to improve their mastery of business and technical English. *Grammar and Style at Work* can operate in conjunction with the 6th Canadian edition of *Technical Communication* or as a stand-alone workbook.

This book contains these self-help components:

- **A Diagnostic Test**. Exercises test your current knowledge and skill levels. You can then use the self-evaluated results to determine which skill areas you most need to improve.
- **Essential Grammar, Usage, and Mechanics**. Each of five exercise sections begins by reviewing basic principles and rules. Each section then allows you to test your skills as you progress.
- **Self-Improvement Exercises**. In three stages, each section of exercises helps you gain command of English grammar, mechanics, and usage. First, practice exercises help you determine how well you have understood the principles discussed in that section. Then, sentence exercises contain errors that you will need to eradicate in workplace writing. Finally, the document critiques will sharpen your evaluation and editing skills. Page 144 provides the standard proofreading marks and correction symbols that you'll need to complete the document critiques.

Starting on page 114, you will find answer keys for the exercises in this workbook.

TABLE OF CONTENTS

TABLE OF CONTENTS

TABLE OF CONTENTS

INTRODUCTION

Often during my four decades of teaching technical communications, I've surveyed business people about the communications skills required for success in their industry or profession. When I've asked what skills and knowledge should be taught in post-secondary English and Communications courses, they've responded with comments such as the following:

- "Our employees need to analyze who reads their correspondence and reports because different readers look for different information and analysis."
- "I like to see short paragraphs and sentences in our company documents, so that they're easy to read and understand."
- "When we hire employees, we look for people who can write clear, concise sentences. We can't afford ambiguity in our instruction manuals and online prompts."
- "We spend more money on communications training than on technical training—it's too bad that our employees don't leave college or university with better communications skills."

The most common response, often accompanied by impassioned, "colourful" language, reads something like this: "Please, please, **PLEASE** make your students pay attention to spelling, punctuation, and grammar! We require correct English in all our documents."

Thriving businesses emphasize language basics because they realize that business success depends on clear, readable, persuasive communications. They also know that they are judged by the quality of the documents they produce, not just by their products and services. Successful business operators know that sloppily written workplace reports, proposals, and correspondence will fail, even if those documents are technically accurate and well designed.

Here's a portion of an actual letter sent by the owner of a construction firm to local businesses affected by upcoming road works:

> T.G. Jones Contracting will be doing Roadworks on 35th Street+36th .Ave starting Sept.17th until Aprx.. Nov.23rdin conjuction with the city of _____ + Rogers head Office. Their will be no parking on 35th. Street, please advice your clients, we will have traffic control, barricades, flashers on the street at all times they must find alternate parking. We will be doing new sewer connections on your properties in which you will be notified 1 day in advance having replacement the same day.

What is your impression of the above document? Based on the quality of the letter, would you be tempted to prejudge the quality of work done by the writer's construction firm? Might such judgments affect your decision to hire such a firm for a future contract? If you manage a business affected by the road and sewer construction, what concerns might you have? And, most of all, do you understand what the writer is trying to convey?

Grammar and Style at Work responds to a need for supplementary material and exercises to help writers improve their skills. This workbook focuses on grammar, sentence structure, vocabulary and usage, technical style, and mechanics such as spelling, punctuation, and numbers. Students can use its self-directed instruction and exercises to improve their mastery of business and technical English. The workbook can operate in conjunction with the 6th Canadian edition of *Technical Communication* or as a stand-alone workbook.

Please understand that this book does not present complete coverage of grammar, usage, and mechanics. For that kind of information and advice, see *Technical Communication*'s "Appendix:

Review of Grammar, Usage, and Mechanics." More exhaustive coverage is available in a handbook such as the *Simon & Schuster Handbook for Writers*, 4th Canadian Edition or the *Prentice Hall Canada Reference Guide to Grammar and Usage*, 2nd edition.

This workbook progresses in three stages:

- On pages 3–7, a **Diagnostic Test** provides exercises that test your current knowledge and skill levels. You can then use the self-evaluated results to determine which skill areas you most need to improve.
- Leading into each group of improvement exercises, a **Review of Basic Principles** provides advice and examples that you will find useful in preparing to complete the practice exercises and the self-improvement exercises.
- The **Self-improvement Exercises** appear in the following order:
    - Basic grammar: parts of speech (nouns, pronouns, verbs, prepositions, conjunctions, interjections, and articles)
    - Sentence grammar
    - Punctuation
    - Mechanics (spelling, capitals, numbers)
    - Vocabulary and style

Following the review of principles in each of the five groups listed above, the exercises help you improve your English grammar, mechanics, and usage. First, a set of practice exercises helps you determine how well you understand those principles and what you need to improve. Next, a more extensive group of improvement exercises contains errors that you will need to avoid or eradicate in workplace writing. Finally, a document critique will sharpen your proofreading and editing skills. Page 144 lists the Correction Symbols that you'll need to complete the document critiques.

Starting on page 114, you will find answer keys for the exercises in this workbook.

Grammar at Work: good grammar is valuable

> "I didn't enjoy studying grammar at school, and I still cringe when I hear that word [grammar], but in my business, our letters, emails, proposals, and bills need to be grammatically perfect and free of spelling errors and punctuation errors. So I guess good grammar is a necessary evil!"
>
> —***Gary Sedlick****, owner and manager of Okanagan Audiolab, which tests hearing at industrial and service locations*

DIAGNOSTIC TEST OF ENGLISH SKILLS

Use the following test to learn whether you need to work on grammar, punctuation, mechanics or vocabulary. After answering the questions, score the test by comparing your answers to the answer key on page 114. Then, use the Skills Assessment form on page 7 to record your number of correct answers for each section.

The following choices apply to items 1 through 10. In each blank, write the letter of the choice that best describes each sentence.

A. incomplete sentence

B. run-on sentence (phrases or clauses are incorrectly strung together)

C. modifying elements misplaced or left dangling

D. structure not parallel

E. nothing wrong

_____ **1.** Stop here.

_____ **2.** This system can recognize voice commands, respond to those commands, and also displaying current settings.

_____ **3.** While the flag rose, the cheering increased.

_____ **4.** Running quietly on a pedestal, I didn't notice the machine at all.

_____ **5.** The printer, an advanced laser jet model, is quite reliable.

_____ **6.** Measuring the stream output, the main function.

_____ **7.** When it's convenient, file your report.

_____ **8.** Several problem areas have been identified, unfortunately, they will all be difficult to solve.

_____ **9.** Running your own business is to face the risk of go bankrupt.

_____ **10.** Smoking is only permitted in designated areas on this campus.

The following choices apply to items 11 through 20. In each blank, write the letter of the choice that identifies the underlined word(s) in each sentence.

A. subject

B. verb

C. object

D. modifier

E. conjunction or preposition

_____ **11.** Run this part over to the loading dock.

_____ **12.** On Thursday, the Leafs lost another game.

_____ **13.** Bombardier Inc. has moved some of its aerospace operations to Northern Ireland.

_____ **14.** Running quietly, the new machine doesn't disturb our conversations.

_____ **15.** The printer is proving quite reliable.

_____ **16.** Measuring electronic output is the main function of this multi-meter.

_____ **17.** When the job's complete, carefully report your findings.

_____ **18.** We have found an assistant for you.

_____ **19.** Your business needs a complete communications audit.

_____ **20.** Smoking is permitted but not encouraged at this job site.

In the blanks for items 21 through 30, write the letter of the word or phrase that properly completes each sentence.

_____ **21.** McDonald's (A. is **or** B. are) experiencing stiff competition. .

_____ **22.** There (A. is **or** B. are) several reasons for the U.S. dollar's slide.

_____ **23.** Every worker who wants to control (A. their **or** B. his or her) own pension plan must complete form W-256.

_____ **24.** Not one of them (A. was **or** B. were) on time for the presentation.

_____ **25.** An important factor affecting that firm's performance is that (A. they try **or** B. it tries) to retain key personnel.

_____ **26.** Once they (A. saw **or** B. seen) the annual report, changes came quickly.

_____ **27.** The number of hooking penalties called in the NHL (A. has **or** B. have) increased since the crackdown on interference.

_____ **28.** If I (A. was **or** B. were) younger, I'd study robotics engineering.

_____ **29.** The receptionist (A. lay **or** B. laid) the drawings on the counter.

_____ **30.** Either Hewlett Packard or Dell (A. has **or** B. have) won the bid.

For items 31 through 40, in the provided blank write the letter of the choice that best describes the corresponding sentence.

A. all punctuation used correctly

B. some punctuation used incorrectly or incorrectly omitted

_____ **31.** McDonald's which is a huge enterprise, is not easy to manage.

_____ **32.** Look for evidence of slippage, shown by deep scratches in the surface.

_____ **33.** Could you complete form W-256 each time an application is successful.

_____ **34.** The project engineer, Susan Marsh, said that, "it's just a matter of time before the bridge fails."

_____ **35.** Cross Canada Engineering has offices in Halifax, Nova Scotia, Ottawa, Ontario, Brandon, Manitoba, and Kelowna, British Columbia.

_____ **36.** After checking his assistant learned that three turbines had failed.

_____ **37.** Do you have a clear idea of the company's "rules of conduct?"

_____ **38.** One main option, then, remains: the turbine must be replaced.

_____ **39.** Not all the spawning beds in Delorme Creek have been contaminated, however, we should continue to monitor the situation.

_____ **40.** Hewlett Packard, Dell, and Acer—these three firms have met the bid requirements.

The following choices apply to items 41 through 50. In each blank, write the letter of the choice that best describes each sentence.

A. error in punctuation

B. error in use of abbreviations or symbols

C. error in use of numbers

D. error in capitalization

E. no errors

_____ **41.** This year's salmon run is high in the Adams river, despite adverse spawning conditions.

_____ **42.** Kovalev, Andrei Kostitsen, & Markov have scored all the team's goals in the past three games.

_____ **43.** Bombardier Inc. has moved some of its aerospace operations to northern Ireland.

_____ **44.** This semester, I'm studying physics, English, and mathematics.

_____ **45.** The printer can print 12 pages per minute, in colour or in black and white.

_____ **46.** The Johnsrude 389B multi-meter—last year's model, has four function ranges.

_____ **47.** When the job's complete carefully report your findings.

_____ **48.** We have found an assistant for you; please explain our operating procedures to him.

_____ **49.** Sale prices of houses in Saskatoon rose by an average of 27% last year.

_____ **50.** Seven smelters will soon appear at this job site, which will necessitate stringent air quality monitoring.

In the blanks for items 51 through 60, write the letter of the word or phrase that properly completes each sentence.

_____ **51.** The McDonald's and Wendy's chains will soon have to modify (A. their **or** B. they're) menus to include low carb items.

_____ **52.** The main (A. effects **or** B. affects) of global cooling will soon appear.

_____ **53.** Our ethics guidelines operate on the (A. principal **or** B. principle) of "no harm, no foul."

_____ **54.** We have four (A. prospective **or** B. perspective) candidates for the vacant position.

_____ **55.** Good team members (A. complement **or** B. compliment) each other's talents; they do not compete for status within the team.

_____ **56.** If we (A. loose **or** B. lose) this contract, our company will have to lay off nearly half of its employees.

_____ **57.** After we hear the last of the six competing proposals, the four members of our committee should discuss the proposals (A. among **or** B. between) ourselves.

_____ **58.** Set the pallets of paving tiles (A. besides **or** B. beside) the brick wall.

_____ **59.** He has advanced the engineering designs (A. farther **or** B. further) than we had thought possible.

_____ **60.** Neither Hewlett Packard or Dell has shown that (A. its **or** B. it's) bid is superior.

Assessment of English Skills

In the space provided below, record the number of questions you answered correctly.

Questions	Number you Got Correct	Skill Area
1–10	_____	Sentence structure
11–20	_____	Grammar: Parts of speech
21–30	_____	Grammar: Verbs and agreement
31–40	_____	Punctuation
41–50	_____	Punctuation and mechanics
51–60	_____	Vocabulary

BASIC GRAMMAR

First, let's define this book's version of the term *grammar*. Often, people use the word to refer to all aspects of writing mechanics—correct use of the parts of speech, sentence structure, spelling, punctuation, capitalization, spacing, verb tense, numbers, and abbreviations. And people often attach emotional (connotative) meanings to the word—"grammar" is often associated with rules, rules, and more rules, many of which don't seem to make sense. That feeling usually results from mechanical teaching and rote learning of the "rules" of grammar. No wonder many of us recoil at the thought of a grammar lesson.

This book looks at grammar in three main ways:

- Grammar should be **descriptive** rather than prescriptive—we should try to understand which word combinations help make our meanings clear; we shouldn't focus on memorizing a host of rules. We should also realize that English is a living, changing language and we need to react accordingly—when enough people use the noun *access* as a verb, we can follow suit. We don't have to write the ungainly phrase "gain access to the library;" we can simply write "access the library." Eventually, the dictionaries catch up to common usage. For example, the 2006 editions of the Oxford English and Merriam-Webster dictionaries added *google* as a transitive verb, following the amazing success of Google Inc.'s search engine.[1]
- In this book, *grammar* refers to **parts of speech** (nouns, pronouns, verbs, adjectives, adverbs, prepositions, conjunctions, interjections) and how they function to convey meaning. Punctuation, mechanics (spelling, capitalization, abbreviation, numbers), and vocabulary usage are treated separately. For example, the difference between *its* and *it's* is considered a spelling/punctuation issue, not a grammar issue.
- In *Grammar and Style at Work*, a separate section called **sentence grammar** displays the structures that make some sentences clearer and easier to read than others.

Also, in this book, the term *grammatical* denotes effective, well-formed phrases and sentences. In other words, *grammatical* reflects a judgment call, a positive evaluation of the way words have been strung together.

Note:

The above distinctions among grammar, sentence grammar, punctuation, mechanics, and word usage help make this book easy to use, because they enable you to focus on one aspect of language at a time. Still, these elements do not operate in isolation, and so this workbook's exercises appear as sentences or paragraphs that test those five components of clear, effective writing.

---

[1] Google's lawyers have actively discouraged using *google* as a verb because, they argue, the practice devalues the Google brand. They don't want their company name to become a generic term like *thermos* or *aspirin*. On October 25, 2006, Google pleaded with the public to "please only use 'Google' when you're actually referring to Google Inc. and our services." ("Do You 'Google'?" posted by Michael Krantz of the Google Blog Team, at http://googleblog.blogspot.com/2006/10do-you-google.html.)

Nouns and Noun Phrases

One of language's most basic functions is to name elements of our experience, whether that experience results from our senses (sight, hearing, and so on) or from our minds—our feelings and our ideas. Therefore, any given dictionary page displays more nouns than verbs or other parts of speech. Some nouns are concrete; they can be experienced directly through our senses (rocks, music, bread, automobile), or they can be identified and measured (kilogram, velocity, calories). Abstract nouns (truth, reconciliation, clarity, element) express our ideas.

Proper and common nouns

All the nouns in the above paragraph are **common nouns** that name general classes. A common noun such as "automobile" is not capitalized. But the name of a particular brand of automobile, such as an Acura RDX, is capitalized. These names of specific items are known as **proper nouns**. Similarly, "music" refers to a general class of things, while a specific song such as "Stairway to Heaven" is capitalized. The sub-classes within a class (rock music or bluegrass or hip hop) are not capitalized.

Nouns as subjects and objects

At the heart of English expression, nouns often denote the person, place, thing, feeling, or idea that forms the **subject** of the sentence. Nouns also function as **objects** of verbs or prepositions. In the following sentence, the nouns are underlined.

> In its corporate <u>blog</u>, <u>Google Inc.</u> asked the <u>public</u> to preserve the <u>value</u> of Google's <u>brand</u>.

The above 16-word sentence uses six nouns, one as a subject (Google Inc.), two as objects (public, value) of verbs (asked, to preserve) and two as objects (blog, brand) of prepositions (in, of). The sixth noun (Google's) appears in the possessive form and essentially acts as an adjective.

Here's a simpler sentence whose subject (engineer) is in a state of being (expressed by the adjective, "uncertain"). This sentence also contains two noun objects (causes, landslide) of prepositions (about, of).

> The <u>engineer</u> is uncertain about the <u>causes</u> of the <u>landslide</u>.

The point here is that each type of word has a job to do, a function to perform. Nouns (and verbs) perform primary functions; English, like all Indo-European languages, absolutely requires nouns and verbs to express basic meaning. The other parts of speech support nouns and verbs—pronouns stand in for nouns; adjectives and adverbs add detail and colour; prepositions, conjunctions, and interjections help complex sentences flow smoothly.

When a noun is not a noun

Let's remember that certain words can function as nouns or as adjectives, or even as verbs:

> These *steel* (adjective) bolts are made of an extremely expensive alloy of *steel* (noun), so *steel* (verb) yourself for the cost of the 1,000 bolts required for the deck supports.

Plural nouns

Nouns can be singular: computer, law, battlefield, textbook. Or they can be plural: computers, laws, battlefields, and textbooks demonstrate the most common way of making a noun plural—we simply add an *s*.

However, we have several other ways of forming plural nouns:

- ox → oxen
- father-in-law → fathers-in-law
- child → children
- medium → media
- life → lives
- box → boxes
- sheep → sheep
- company → companies
- woman → women
- datum or data → data
- cactus → cacti
- criterion → criteria
- mouse → mice
- crisis → crises
- appendix → appendixes or appendices

Why does English have so many ways of forming plural nouns? The short answer is that English has been forged from one root language (Anglo-Saxon, an early form of German) and four modifying languages (French, Latin, Greek, and Norse). So, for example, the plural form of *goose* is *geese* because those words came from the original German, but by the time the words *caboose* and *moose* came into English, plurals were being formed by adding an s sound, or in the case of *moose*, with no change in the word. Words that have come from Latin (*medium*, *datum*) and Greek (*crisis*, *criterion*) brought their native plural forms with them.

Possessive nouns

Most English nouns add *'s* to denote ownership:

Steven Carr's Jaguar   my sister's apartment

However, the "ownership" doesn't have to be legal:

the van's motor each worker's locker   a dollar's worth

Here are the usual conventions for using apostrophes to indicate possession:

- For singular nouns, including nouns ending in *s*, acronyms, and indefinite pronouns, add *'s*:

  Mark's laptop   Chris's laptop   OPEC's oil revenues   everyone's answers

- For plural nouns ending in *s*, add only the apostrophe:

  the Joneses' trust fund three players' equipment

- For plural nouns not ending in s, add *'s*:

  women's opportunities men's privileges

- For compounds, add *'s* only to the last word:

  someone else's idea my mother-in-law's advice the Prime Minister's veto

- To indicate individual ownership, add *'s* to each name:

  the engineer's and the technologist's drawings Tom's and Rhianna's desks

- To indicate joint ownership, add *'s* only to the last name or to each name:

  Tom and Rhianna's car Tom's and Rhianna's car

Noun phrases

A noun phrase contains a noun and one or more modifiers:

the old bypass road several responsible biologists the woefully late report

A noun phrase might appear as a subject:

The old bypass road should be upgraded to handle the construction traffic.

It might function as an object:

Jeannine silently slid *the woefully late report* under her professor's door.

Or it might function as a complement:

Norma Cross, the lab director, is one of the *several responsible biologists* whose reputations have been tarnished by allegations of careless sampling methods.

Practice session—nouns

Underline the preferred choice within each set of parentheses in the following sentences. You will find answers for these exercises on page 114.

1. Our firm's (*specialtys*, *specialties*) are in data gathering and statistical analysis.
2. Water temperature exceeded 34 degrees (*celsius*, *Celsius*) for the second day in a row.
3. Place the heavier equipment on the lower (*shelfs*, *shelves*).
4. Send your company newsletter items to the (*editor-in-chief's*, *editor's-in-chief*, *editor-in-chiefs*) email address.
5. Our shop guarantees its repairs to (*RVs and boats*, *RV's and boat's*, *RVs and boat's*) for three months.

6. Despite his heavy work load this spring, (*Ryan Collins'*, *Ryan Collin's*, *Ryan Collinses*, *Ryan Collins's*) reports have all been submitted on time.

7. George McAvoy, the acting (*pit boss*, *Pit Boss*, *pit Boss*), entered the mine just as the support beams collapsed.

8. (*Bill and Wanda's*, *Bill's and Wanda's*) reports recounted quite different versions of the meeting with the city's solicitors.

9. A thorough analysis of the stream flow (*data*, *datas*) showed very little evidence to support the conservation society's accusations.

10. Our company has been in business in Fredericton since the early (*1980s*, *1980's*, *1980,s*).

Improvement exercises—nouns

In items 1 through 5, underline the common nouns and circle the proper nouns. You will find answers for these exercises on page 114.

1. Each of the consultants wrote a report about Riverbend Dam's projected cost overruns.

2. In this book, "grammatical" refers to effective, well-formed phrases and sentences.

3. Grammar studies the rules governing the use of language, as Dr. Benson has stated many times.

4. Written technical communication is not the only subject covered in *Technical Communication*, a book designed for present and future technical communicators.

5. Brick walls can withstand considerably higher temperatures than concrete board walls.

In items 6 through 10, underline the subjects and circle the objects. You will find answers for these exercises on page 115.

6. Each of the consultants wrote a report about Riverbend Dam's projected cost overruns.
7. In this book, the term "grammatical" refers to effective, well-formed phrases and sentences.
8. Grammar studies the rules governing the use of language, as Dr. Benson has stated many times.
9. Written technical communication is not the only subject covered in *Technical Communication*, a book designed for present and future technical communicators.
10. Brick walls can withstand considerably higher temperatures than concrete board walls.

In items 11 through 20, underline inappropriate noun plurals and possessives and write the correct form(s) in the space provided. You will find answers for these exercises on page 115.

11. Rollover rates for SUVs are nearly twice that of sedans, and 30 percent higher than that of pickup truck's.
12. Please pay attention to all the criterions listed in the test procedure bulletins.
13. Sundins career as a Toronto Maple Leaf appears to be ending.
14. His partner's offers were insufficient to resolve the companies debt load.
15. Plagiarism may be more prevalent than ever, but so is detection softwares.
16. Revenue Canada requires more stringent accounting procedure than our firm can afford.
17. According to Rene Corbeils report, the bridge is no longer safe.
18. Every one of the seven stimuluses introduced in the test sequence provoked abnormally high levels of radiation.
19. Don't expect your cellular phone's to receive strong signal's in the far northern Arctic.
20. Who knows which shoreline's will be most affected by Global warming?

Document critique—nouns

The following document may contain errors in grammar, capitalization, punctuation, or spelling. Underline all errors and identify them using standard proofreading marks. (See the Correction Symbols list on page 144.) You will find answers for this exercise on page 115.

MEMO

To:	Harlan Devonshire, Field Supervisor
From:	Arjun Atwal, Accounting Office
Date:	June 24, 2009
Subject:	Expense Claims

We have had difficultys establishing the accurcy of recent expense claims submitted by your exploration group. Especially this past month.

Would you please review all your team members expense claims before they are sent to are office? Im sure that you no when claims are excessive and you'll be able to correct them before we have to deal with them.

As an example of what I mean, I've attached Rose Francis' claim for this May. I find it hard to belief that she could spend $900 on camp supply's in addition to the food allowance!

Pronouns

In order to create smooth phrasing and to avoid repeating nouns, writers (and speakers) use pronouns.

> Each Monday morning, the department heads report the previous week's production in their sections of the plant; *they* fret about these reports because *their* bonus pay is based on these production figures.

The pronouns *they* and *their* stand in for the noun *heads*, which is the antecedent of the pronouns; heads is the antecedent of *they* and *their*. When the antecedent is plural, the replacement pronoun(s) must be plural, to avoid confusing the reader. However, when the antecedent is singular, the replacement pronoun must be singular:

> When Canwood Furniture announced *its* closure, 91 employees lost *their* jobs.

In the above example, it's clear that Canwood Furniture is a single entity, so the pronouns *it* or *its* logically substitute for that proper noun. The plural pronoun *their* substitutes for the plural noun *employees*.

However, the issue may not be so clear with a noun phrase such as the *Montreal Canadiens*. At first glance, *Canadiens* looks like a plural word and if one is thinking of the many people in the Canadiens organization, the correct pronoun replacements would be *they* or *their*. On the other hand, if one is thinking of the organization itself, the logical pronouns would be *it* or *its*. Just to be sure about that point, let's consider these examples:

> The Canadiens may finish in top spot in the league for the first time in many years, as *they* rely on *their* young players to provide scoring punch. The team has continued to be financially successful; now *its* on-ice performance matches *its* corporate performance.

Plural antecedents

A pronoun might have more than one antecedent:

> Ricky Ruiz and Elma deLoos had their laptops stolen from their lockers on Wednesday.

Each of *Ricky Ruiz* and *Elma deLoos* would be a singular antecedent, if used alone. However, in the above sentence they form a plural antecedent, so the plural pronoun must be used.

Unclear antecedents

Sometimes, a pronoun's antecedent is unclear:

> Elma deLoos told Marcie Harper about the change to *her* work schedule; *she* was unhappy about the changes, because *she* had already arranged for babysitting.

How would a reader understand the references to *her* and *she* in the above sentence, unless the reader already knows the situation? The acronym COIK has been coined for such unclear references—"clear only if known!"

Such a sentence can be made clear by repeating nouns where necessary:

> Elma deLoos told Marcie Harper about the change to *Marcie's* work schedule; *Marcie* was unhappy about the changes, because *she* had already arranged for babysitting.

Or,

Elma deLoos told Marcie Harper about the change to *Elma's* work schedule; *Elma* was unhappy about the changes, because *she* had already arranged for babysitting.

Pronoun case

Grammar handbooks normally identify three pronoun *cases*, the term used for forms in which pronouns are employed: subject, object, possessive. To those three, we could also add the reflexive case, because increasingly the pronoun *myself* is mistakenly used instead of *I* and *me*. The following table lists the singular and plural forms of these cases.

Table 1 Pronoun Cases

	Subjective		Objective		Possessive		Reflexive	
	Singular	*Plural*	*Singular*	*Plural*	*Singular*	*Plural*	*Singular*	*Plural*
First person	I	we	me	us	my, mine	our, ours	myself	ourselves
Second person	you	you	you	you	your, yours	your, yours	yourself	yourselves
Third person	he	they	him	them	his	their, theirs	himself	themselves
	she	they	her	them	her, hers	their, theirs	herself	themselves
	it	they	it	them	it, its	their, theirs	itself	themselves

Notes:

- The *subjective* case is sometimes called the *nominative* case.
- Later, this section discusses the subjective, objective, and possessive relative pronouns *who*, *whom*, *whoever*, *whomever*, and *whose*.

Subjective pronouns perform the action or precede a statement of being:

He ran the samples through a spectrometer. *She* is happy to be here.

Subjective pronouns can also follow a linking verb, as the following two examples demonstrate:

Are you certain that it will be *she* who is assigned to the Samuel Plains project?

Who is there? It is *I*.

It can be tricky to choose whether to use the subjective or objective case when words are omitted in comparative statements.

She is more intelligent than *I*. [The omitted words here are "am intelligent."]

Professor Magellan likes you more than *me*. [The omitted words are "more than he likes."]

In the first case above, a subject is needed—"*I* am intelligent."

The second sentence requires an object—"… more than he likes *me*."

Problems with pronouns

In many situations, you will know whether to use the objective, subjective, or possessive form of a pronoun, but some pronoun problems do plague writers (and readers).

1. *Who* and *whom* versus *that*

Recently, speakers (and writers, too) have been using *that* instead of *who* or *whom*, perhaps because many people don't know that *who* operates in the subjective case and *whom* is designed to receive action (objective case). Or perhaps many people are just lazy and they don't want to think about the distinction. Or more likely, we simply mimic what we hear, without thinking about it much. However, careful writers should use *that* to refer to inanimate objects, animals that are not named, and ideas.

Remember that *who* and *whom* are for humans or for animals that have been given names.[2]

> The young man *who* retrieved the record-breaking home run ball *that* Barry Bonds hit has offered to sell the ball; he believes that the "finders/keepers" rule entitles him to the cash. Others say that the baseball belongs in the Major League Baseball hall of fame in Cooperstown. *Whom* do you support in this controversy? And what if Dusty the Bloodhound, *who* hangs out regularly at TriCom Park, had "fetched" the ball? Should Dusty be able to trade the ball for a 20-year supply of Dog Chow and a plush dog house?

But how do you tell when to use *who* and when to use *whom*? The down-and-dirty technique identifies the key clause and then decides whether it needs a subject, which would equate to *who*, or an object that would require *whom*. For example, the clause "Who/Whom should be Prime Minister?" requires a subject—if you substituted *he* or *him* for *who* or *whom*, you would write, "**He** should be Prime Minister." In other words, the correct version of the above example would be "**Who** should be Prime Minister?"

How about this example? "Who/Whom do you want to lead the country?" In this case, the key thought is "Do you want _____?" The clause requires an object such as *him* (Do you want **him**?), so the correct version of the above example would be "**Whom** do you want to lead the country?"

Additional examples follow:

- Who/whom saw Elvis at the car wash? (**he** saw Elvis)
- The professor glowered at whoever/whomever spoke out of turn. (**he** spoke out of turn)
- Who/whom did you see at the rally? (you saw **him**)
- I will hire whoever/whomever the hiring committee recommends. (the committee recommends **him**)
- I will hire the person who/whom is best qualified. (**he** is best qualified)

---

[2] Mignon Fogarty, a.k.a. "The Grammar Girl," who speaks entertainingly about grammar, prefers the traditional rule outlined above. However, she points out that "it's more of a gray area than some people think, and if you have strong feelings about it, you could make an argument for using *that* when you're talking about people. But my guess is that people who use *who* and *that* interchangeably do it because they don't know the difference." See the transcript of her podcast Episode 24, posted November 4, 2006 at http://grammar.quickanddirtytips.com/who-versus-that.aspx.

2. Unclear reference pronouns

To help readers understand the relationship between a pronoun and the noun it replaces, make sure that your pronoun refers to a noun that is mentioned nearby and not merely implied. To further help your reader, place that pronoun as close as possible to the noun it replaces.

> Unclear reference: Jason gave Roger his keys to the company van. [Are these Jason's keys or Roger's keys?]
>
> Clear reference: Jason gave his keys to the company van to Roger. [The possessive pronoun his is now closer to the noun it references.]

Also, don't use the vague *it* or *they* if such words do not refer to a specific word or phrase. In some cases, you will need to reword a sentence in order to include the noun or phrase you had in mind.

> Unclear: Our office recycles as much paper as possible because *it* helps the environment. [Does *it* refer to the paper, or to the office, or to something that's not even in the sentence?]
>
> Clear: Our office recycles as much paper as possible because *such efforts* help the environment.
>
> Unclear: When we asked the company whether *they* would follow the new guidelines for capturing plant emissions, *they* did not respond. [Who are *they*? And does the writer have persons in mind for each of the two *they*s?]
>
> Clear: When we asked the company whether *it* would follow the new guidelines for capturing plant emissions, *its Public Relations Office* did not respond.

Remember:

The pronouns *this*, *that*, *these*, and *those* provide clear references only when they are immediately followed by a noun or a noun phrase.

> Unclear: Negotiators announced an expanded health plan, a reduction in dental coverage, and on-site child-care facilities. *This* caused employee protests. [Does *this* refer to the health plan, the reduced dental coverage, or the child-care facilities?]
>
> Clear: Negotiators announced an expanded health plan, a reduction in dental coverage, and on-site child-care facilities. *This* reduction in dental coverage caused employee protests.
>
> Unclear: The beach sand has eroded away, leaving a rocky beach. *This* has made the property less desirable for tourist campers.
>
> [Does *this* refer to the erosion, or to the rocky beach, or to the cause-effect relationship?] The three following edits expresses what the writer really meant to say.
>
> Clear: The beach sand has eroded away, leaving a rocky beach. *This erosion* has made the property less desirable for tourist campers.
>
> Clear: The beach sand has eroded away, leaving a rocky beach. *This property* is now less desirable for tourist campers.
>
> Clear: The beach sand has eroded away, leaving a rocky beach, *which* has made the property less desirable for tourist campers. [*Which* refers to everything that precedes it.]

3. Misuse of reflexive pronouns

Reflexive pronouns, which are sometimes called *intensive* pronouns, reflect or intensify the nouns or pronouns they invoke:

> I finished it *myself*. Give *yourself* some credit! They have only *themselves* to blame.

However, they should not be used as subjects:

> No: *Myself* and three others spoke at the meeting.
>
> Yes: Three others and *I* spoke at the meeting.

Nor should they be used as objects:

> No: The credit was given to *myself* and Janus Czorzinski.
>
> Yes: The credit was given to Janus Czorzinski and *me*.

4. Lack of pronoun agreement

Pronouns should agree in number, person, and gender with their antecedents:

> The students handed in their assignments, confident that *their* work would be graded fairly.
>
> For the third day in a row, Grant called in to say that *he* was sick.

The above two examples are quite straightforward, but compound antecedents pose more of a challenge. So, when two parts of a compound antecedent are joined by *and*, use a plural pronoun:

> Television and radio commercials often express *their* pitches in colloquial language.

A compound antecedent that is preceded by *each* or *every* requires a singular pronoun:

> Every period, comma, colon, and semicolon has *its* useful function.

When the parts of a compound antecedent are joined by *or* or *nor*, the pronoun agrees with the nearest antecedent:

> Neither the company nor the union members would revise *their* bargaining positions.

5. Possessive pronouns

Remember that not one of the possessive pronouns requires an apostrophe. Not one of them!

Let's repeat that point—*its*, *his*, *hers*, *my*, *mine*, *our*, *ours*, *their*, *theirs*, *your*, *yours*, *whose*—none of these pronouns uses an apostrophe. The apostrophe is used with nouns to indicate possession.

6. The relative pronouns *that* and *which*

The issue of whether to use *that* or *which* has plagued writers for generations, but luckily there's a straightforward rule of thumb: *that* introduces essential phrases or clauses, while *which* leads into nonessential phrases and clauses.

> The car *that* he sold to me is in better shape than I thought. [The *that* clause in this sentence is essential to the meaning.]
>
> The Honda Civic, *which* has been the best-selling car in Canada for the past three years, suits my transportation needs perfectly. [The *which* clause is not essential to the meaning; this clause presents extra information.]

7. Generic pronouns

English has masculine pronouns (*he*, *him*, *his*) and feminine pronouns (*she*, *her*, *hers*), but the language doesn't have gender-neutral singular pronouns. Until recently, English speakers used masculine forms for either masculine or feminine expressions. In a post-feminist world, that usage is deemed to be a gender bias and therefore unacceptable. In the 1970s, there was a concerted effort to introduce the words *tey*, *tem*, and *ter* as gender-neutral pronouns, but that initiative failed, as have previous attempts to introduce generic singular pronouns.

To avoid sexist language, we need some way of dealing with sentences such as the following:

> A student will succeed in a writing class only if *he* completes all *his* assignments.

A writer might be tempted to use the words *she* and *her*, perhaps in an attempt to compensate for centuries of male sexism:

> A student will succeed in a writing class only if *she* completes all *her* assignments.

However, that solution is no better than the male version, is it? Many people now hijack the plural forms, *they*, *them*, and *their*, but that solution has its limitations, also:

> A student will succeed in a writing class only if *they* complete all *their* assignments.

Apparently, that single student has been cloned! *They* and *their* do not agree in number with their antecedent, *a student*. Still, ever-increasing usage of this illogical approach may soon crown this solution as the winner. Here's one advocate for the practice:

> I will state for the record that I am a firm believer that someday *they* will be the acceptable choice for this situation. English currently lacks a word that fits the bill, and many people are already either mistakenly or purposefully using *they* as a singular generic personal pronoun; so it seems logical that rules will eventually move in that direction. (Mignon Fogarty, transcript of podcast Episode 29 posted December 8, 2006 at http://grammar.quickanddirtytips.com/he-they-generic-personal-pronoun.aspx)

Here are some other ways to use a pronoun to refer to an unknown gender or to refer to both males and females:

- Use the awkward combinations of *he or she* or *his or hers*

  A student will succeed in a writing class only if *he or she* completes all *his or her* assignments.

- Recast the antecedents and pronouns into the plural form:

  Students will succeed in a writing class only if *they* complete all *their* assignments.

- Reword the sentence to avoid the problem altogether:

  Success in a writing class depends on completing all the assignments.

  In most cases, rewording is not only possible, but preferable.

Practice session—pronouns

Underline the preferred choice within each set of parentheses in the following sentences. You will find answers for these exercises on page 116.

1. Return that spectrometer to Gordon Burns or (*myself*, *me*) when you've finished using it.
2. Despite the strong Canadian dollar and low U.S. demand for lumber, Foreman Brothers Ltd. is increasing (*their*, *its*) production of framing lumber.
3. He is a basketball player (*that*, *who*) can rebound, run the floor, and shoot three-balls.
4. Everyone has (*his or her*, *their*) email address posted on the class message board.
5. Our shop guarantees (*its*, *their*) repairs for three months.

6. After the environmental assessment board tabled (*its*, *their*) report, three mining firms disputed the board's conclusions.
7. When you challenged the security officer's dispersal order after the pep rally, did you know to (*who*, *whom*) you were speaking?
8. (*Myself*, *I*) and Toni Colette were the first observers to recognize the SARS symptoms.
9. Let's be honest—(*who*, *whom*) knows how global warming will affect the East Coast?
10. Every collector bottle, pipette, and Petri dish has (*its*, *their*) function in this experiment.

Improvement exercises—pronouns

Below, replace the underlined nouns with the correct pronouns. (Answers are shown on page 116.)

1. Each of the consultants wrote a report about Riverbend Dam's projected cost overruns.
2. To which government agent will you send the gas well report?
3. The Environmental Protection Board has issued the Environmental Protection Board's assessment of Alberta's gas-powered power generators.
4. After examining the project files closely, can you tell which person's drafting errors contributed to the failed earth dam?
5. The engineering firm submitted the firm's proposal three days early.

In each of items 6 through 10, write the correct pronoun in the space provided. You will find answers for these exercises on page 116.

6. Each of the consultants wrote a report about _____ (his or her, their) view of Riverbend Dam's projected cost overruns.
7. Make sure to include _____ (*you're*, *your*) receipts when you submit this month's travel expense claim.
8. _____ (*We*, *Us*) technologists know what will work, not the project engineers.
9. Bring the design to _____ (*myself*, *me*) for final approval.
10. Much of the time, our _____ (*company's*, *companies*) equipment is inadequately maintained.

The following 15 sentences may contain errors in pronoun usage, spelling, or punctuation. Rewrite those sentences, correcting the errors. If a sentence is already correct, write ***C*** in the space following the sentence. You will find answers for these exercises on pages 116 and 117.

11. Myself and all the team at EBS look forward to hearing from you.
12. These dense patches of forest are not a vulnerable ecosystem, it is in generally good condition, but they are found on steep slopes and would be best left undisturbed by development.
13. In order to create a new water intake at Lambeau lake, a new structure will have to be built below the current water level of 112 feet. In order to accomplish this, the lake level has to be dropped to allow construction at an elevation of about 92 feet.
14. The housing used to protect the stream meters is a 2-inch galvanized pipe, that rests flush with the riprap boulders on the left bank.
15. The system includes two cigar-sized dataloggers which contain both pressure transducers and temperature recorders. These are housed in protective casings.
16. Canadian National Railways (CN) has mapped the location of all known streams that are crossed by their railways, on a GIS-based railway Information System.
17. Box type concrete culverts were constructed with a concrete skirt (Figure 15) downstream of the outlet. The reason for this was to reduce scour below the skirt, which would block fish passage.
18. To transfer vinyl records to MP3 files, it must meet certain specifications.
19. Just between you and I, our network specifications will not meet Microsoft's standards.
20. Martin Technologies has just announced who they have chosen as their main supplier of cable and peripherals.
21. Its time for the company to review all its' recruiting practices.
22. Each of Frances and Donna have told their clients about the new line of excavators.
23. Austin St. Pierre of St. Johns Ambulance Services will explain their new billing policy.
24. In Calgary. officials ended their (intersection) red light camera program after statistics showed a 40 percent drop in ticketable offenses.
25. I found that in researching this report it was useful in learning about management techniques.

Document critique—pronouns

The following document may contain errors in grammar, capitalization, punctuation, or spelling. Underline all errors and identify them using standard proofreading marks. (See the Correction Symbols list on page 144.) You will find answers for this exercise on page 117.

Concept Architectural Designs

MEMO

To:	Devin Marchand, Architectural Technologist
From:	Shirley Tostenson, Managing Partner
Date:	April 16, 2009
Subject:	Preliminary Research for Potential project

We are in the preliminary stages of preparing a proposal to design Halifaxes potential Seaport Redevelopment project. I am assigning yourself to conduct reserch that will place our proposal in context.

Please review all the available literature on the folowing group of Waterfront developments, Granville island, Vancouver, the Distillery district, Toronto, the custard Factory, Birmingham, England, the forks, Winnipeg, make note of its capitalization, scope, end uses, community involvement, sources of capital, architectural themes, blend of commercial, retail, arts and crafts and residential.

Complete the survey by April 25 and bring the results directly to myself. This is a high priority so drop everything else which you are doing. Don't discuss this outside of the office, we don't this leaking out to who might be also submitting a proposal.

Verbs

Verbs describe action (active verbs):

> The department heads *report* the previous week's production.

And verbs express a state of being (linking verbs):

> My feet *are* my only mode of transportation. [Bob Marley used more colourful, albeit ungrammatical, phrasing when he sang "My feets is my only carriage."]

Here are some of the more common active verbs you'll use in technical writing:

observe	report	perform	measure	conclude	recommend
contain	develop	agree	reduce	use	include

Nearly all linking verbs are forms of *to be*:

am, is, are, was, were, will be, have been

However, *sounds*, *seems*, *looks*, and *appears* are sometimes used to express a state of being:

This winter *seems* like the worst one we've had in the past 20 years.

He *looks* tired. You *look* tired.

The road *appears* to have been repaired since last month's floods.

Their praise of employee performance *sounds* insincere.

To be and the other four linking verbs connect the words that precede them with what follows in the sentence. That link provides a useful function in many sentences, but sets up semantic pitfalls. For example, a statement such as "She is the best writer on the research team" contains two potential problems:

1. Although it appears to state a fact, it really only expresses an opinion, and in this case the opinion is not easy to verify.
2. Using the present tense of the linking verb sets up a static judgment. Statements that feature *is* or *are* lead to the assumption that people are consistent and unchanging. However, "she" may not always submit the best pieces of writing!

Linking verbs often contribute to another phrasing problem—their presence can result in wordy phrasing. See the "Vocabulary and Style" chapter of this workbook for advice on how to reduce wordiness by replacing linking verbs with active verbs.

Verb Tenses

The concept of past, present, and future time is built into English:

Past: We *reported* that there's been no change in caribou migration patterns for the past 12 years.

Present: They need to *report* the findings concerning caribou herd enhancement.

Future: We *will report* current methods of identifying and culling the herd's disease carriers.

The past, present, and future simple verb tenses have predictable forms. The regular verbs (which constitute the majority of all verbs) end the past tense in *ed* while the future tense verbs are preceded by *will* or *shall*. Present tense forms can be slightly more complex, as Table 2 shows.

Table 2 Present Tense Verb Forms

	First Person	Second Person	Third Person
Singular	I report	you report	he/she/it reports
Plural	we report	you report	they report

The verb's root form, *report*, adds an *s* when preceded by *he*, *she*, or *it*.

Here are other forms of a regular verb such as *report*:

Table 3 Additional Verb Tenses

	First Person	Second Person	Third Person
Present perfect	I/we have reported	you have reported	they have reported
Past perfect	I/we had reported	you had reported	they had reported
Future perfect	I/we will have reported	you will have reported	they will have reported

When actions occur at the same time, keep verbs in the same tense:

Each time the protestors *moved* forward, the riot police *raised* their shields.

The project manager *has reported* that he *has found* safety infractions.

However, when actions occur at different times, change case accordingly:

The project manager *reported* that he *has fired* two workers for safety infractions.

If the tests *show* abnormalities, they *will confirm* our suspicions that the cattle *have been ingesting* tainted feed.

Irregular verbs

Not all verbs follow the patterns shown above. Here's a sampling of irregular verbs:

Present	Past	Past Participle
run	ran	run
become	became	become
go	went	gone
grow	grew	grown
make	made	made
take	took	taken

The most irregular of all verbs is *to be*:

Table 4 Tense Forms of *To be*

	Singular	Plural
Present	I am/you are/he is/she is/it is	we are/you are/they are
Past	I was/you were/he was/she was/it was	we were/you were/they were
Future	I will be/you will be/he, she, it will be	we/you/they will be
Present perfect	I have been/you have been he/she/it has been	we/you/they have been
Past perfect	I /you/he/she/it had been	we/you/they had been
Future perfect	I/you/he/she/it will have been	we/you/they will have been

Transitive and intransitive verbs

Transitive verbs transfer action to an object; intransitive verbs express an action, but do not transfer it to an object. Here are some examples that show three sets of verbs (*lie/lay*, *sit/set*, and *rise/raise*) that often pose problems for writers:

Table 5 Troublesome Verbs

Intransitive	Transitive
Even though the Varsity Blues team has lost 12 games in a row, don't expect the Blues to *lie* down and let the Mustangs roll over them.	Just *lay* that rebar beside the driveway.
Don't *sit* in the General Manager's chair!	Please *set* the cases of beer on the floor behind the bar.
We can expect the Bombardier's stock to *rise* quickly, now that the company has won the bid to supply all those trains to France.	When you *raise* the issue of tainted water, I will call for a full inquiry.

Remember that *lie* and *lay* are different verbs, but the present tense of *lay* and the past tense of *lie* are spelled and pronounced the same.

Present	Past	Past Participle
I lie	I lay	I have lain
I lay (something down)	I laid (something down)	I have laid (something down)

To further confuse matters, an alternate meaning for *to lie* (to tell an untruth), uses *lied* as its past tense and past participle forms!

Subject-verb agreement

Clear phasing often depends on agreement between subject and verb, in person (first, second, third) and in number (single or plural). Simple clauses like "my watch keeps perfect time" present few difficulties, but here's a more complex sentence that might prove vexing:

> Symptoms of lymphocytic bronchitis include shortness of breath.

Although the singular word *bronchitis* is close to the verb and may seems to be the subject of the sentence, the actual subject, *symptoms*, is plural and it requires a corresponding plural verb.

In cases where the subject may be singular or plural, you need to examine the surrounding sentence to decide which verb form to use. In the following examples, the subject is underlined and the verb is italicized. Explanations appear in brackets.

> Almost <u>all</u> of the paragraphs, especially the first one, *require* a clearer topic sentence. [The plural subject requires a plural verb; *almost* and *of the paragraphs* modify *all*.]
>
> Either the <u>Director</u> or the research team <u>members</u> *have* issues that must be resolved. [When subject words are joined by *either … or*, *neither … nor*, or *not only … but also*, the verb agrees with the closer subject.]
>
> <u>Running a marathon</u> *requires* tremendous willpower. [When a whole clause or phrase is the subject, use a singular verb.]
>
> Each <u>speaker</u> must *talk* no longer than 10 minutes.
>
> <u>Each</u> of the computers *has* a serial number engraved on the back wall of its chassis.

[Indefinite words such as *every*, *each*, and *any* require a singular verb when they act as a subject or precede a subject word.]

McCains *is* hiring plant workers for the summer months.

Surprisingly, *Seven Canadian Truisms has* higher sales figures in the U.S. than in Canada.

"Felicitations" *is* a phrase he often uses when greeting people.

[Names of companies, titles of books, and words used as terms require a singular subject.]

Twenty spelling errors *is* excessive in a five-page document.

Did you know that 20 scientists *are planning* to move to the U of S research facility?

[Quantities that refer to a total amount require a singular verb, while quantities that refer to individual, countable units require plural verbs.]

The staff *prefer* to eat lunch away from the office. [When you perceive an apparently singular subject as a countable collection of individuals, use a plural verb.]

The log *book* and the test *report* are in my office. [A compound subject requires a plural verb.]

The news media *are* very interested in the global warming issue. [A plural subject requires a plural verb.]

Active and passive voice

In the active voice, the subject performs the action of the verb. In passive voice, the subject receives the action.

Passive: The report was written by me.

The experiment was conducted by our group.

The project has been completed.

[Weak and wordy]

Active: I wrote the report.

Our group conducted the experiment.

The field engineers have completed the project.

[Stronger and more concise]

Passive: All doors in this building will be locked by 8 p.m.

Active: The commissionaire will lock all doors in this building by 8 p.m. [Change in emphasis, change in responsibility]

Passive: While conducting these experiments, the rats were seen to become excited when food was brought in. [Passive voice helps contribute to the dangling modifier]

Although the passive voice usually leads to wordy, less powerful phrasing than the active voice, the passive voice does have its uses:

- Passive voice can place emphasis on the action, rather than on the doer of the action.

  The project has been completed on time and on budget.

- Sometimes a writer doesn't know, or doesn't want to say, who performed the action.

  Seven of the golf course's greens were damaged by vandals on Tuesday evening.

The project was completed three weeks behind schedule.

Practice session—verbs

Underline the preferred choice within each set of parentheses in the following sentences. You will find answers for these exercises on page 117.

1. One hundred dollars (*is*, *are*) far too much to pay for a DVD.
2. Mary, Barry, and Larry (*has*, *have*) gone to look at computer programs.
3. Which one of those guys (*sings*, *sing*) in the choir?
4. *No Alternatives* (*was*, *were*) very tense reading.
5. One of the boys (*was*, *were*) supposed to meet you at the train station.
6. Every one of these buildings (*need*, *needs*) a new roof.
7. Everyone has (*his or her*, *their*) email address posted on the class message board.
8. The measles (*was*, *were*) thought to be eradicated until new cases developed in the 1990s.
9. Recommendations for mitigating potential impacts (*are*, *were*) presented.
10. Someone has inadvertently (*lain*, *laid*) the specimen case on the van's roof.

Improvement exercises—verbs

In items 1 through 5, rewrite the sentences so that they use the active voice. **However**, if you believe a sentence should be left in the passive voice, say so, and then explain your decision.

You will find answers for these exercises on pages 117 and 118.

1. The experiment was conducted by our group.
2. A final report will be prepared upon receipt of your comments.
3. A list of upcoming squash games has been posted on the bulletin board.
4. Following a storm in March 2008, woody debris, boulders, and gravel were removed from the site.
5. The firm's proposal was submitted by the engineering firm three days early.

In items 6 through 10, replace linking verbs with active verbs.

6. The following report is a summary of the project's progress to date.

7. Also included in the report are comments intended to give an estimate of ecosystem richness and the presence of invasive alien species.

8. Successive approximations is a method of converting analogue to digital signals, in which comparing the unknown input against a known input produced by the D/A converter.

9. Their software is used by or firm for tracking traffic on our website.

10. Seeking reliable data on who utilizes open source software, Freeware.com provides the best option.

The following 18 sentences may contain errors in subject-verb agreement or in verb tense. Or linking verbs might help bury the action in nouns or adjectives. Rewrite sentences if necessary, correcting grammatical and/or stylistic errors. If a sentence needs no improvement, write ***C*** in the space following the sentence. You will find answers for these exercises on page 118.

11. In March 2009, Northern Nurseries and Landscaping began their reconstruction of the banks of MacDonald Creek.
12. A new crib wall construction was done in July 2008.
13. It is my opinion that online gambling is a threat to small local casinos.
14. The test method and the answer key is faulty.
15. The system is two cigar-sized dataloggers, each of which is containing a pressure transducer and a temperature recorder. These are housed in protective casings.
16. As instructor of this course, it is my role to be facilitative and provide learning opportunities.
17. I will be graduating from George Brown College's Health Informatics program on May 26.
18. No work program was plan for at the time of this environmental assessment.
19. By allowing the engine to breathe easier typically is producing 15 more horsepower.
20. The structure of the following analysis is a combination of an assessment, a comparison, and an eventual recommendation for doubling next year's research budget.
21. Every time I suggested an alternative plan, my boss was raising objections of a budget nature.
22. The rough draft of the guidance document is recommending that pharmaceuticals would be labelled if they would pose an environmental risk when released into the water supply.
23. He must have grew four inches over the summer holidays!
24. The Accident Prevention Board like to hold informal sessions that include all stakeholders.
25. The Accident Prevention Board is comprised of seven members from a wide range of industrial occupations.
26. There (is, are) some oranges in the fruit bowl.

27. Each of the winners receive $500 and their choice of Corel software.
28. The media has its own agenda regarding the Kyoto Accord.

Document critique—verbs

The following document may contain errors in grammar, capitalization, punctuation, or spelling. Underline all errors and label them using standard proofreading marks. (See the Correction Symbols list on page 144.) You will find answers for this exercise on pages 118 and 119.

Concept Architectural Designs

MEMO

To:	Shirley Tostenson, Managing Partner
From:	Devin Marchand, Architectural Technologist
Date:	April 22, 2009
Subject:	Preliminary Research for Potential project

As you have request, I have been engaging in research of several waterfront developments. As you will see in the attached research Overview, some very exciting developments have happen in canada. England, too.

I guess you should be knowing where I found the information about the respective water front developments. Well, I have got a lot of info from the internet, but most of it is being supply by the development corporation themself. So, its suspect. That's why I then went to newspaper archives for the Vancouver, Toronto, Birmingham, and winnipeg daily Papers. Especially the bisiness sections.

Also, I was able to get documents from the planning offices in the city of Vancouver and also the city of winnipeg. I had a friend in City hall in Toronto, so she is able to give me background details of the proposals submitted to the City for the Distillery District development, that I received last week. I have phoned the City CLERK in Birmingham, he promised to send a package of materials which will have described the whole Custard Factory development, complete with construction costs. I hope it come because as you see there was not much else available.

Thanks for having been assigning myself this research project which had been very interesting.

Adjectives and Adverbs

Adjectives and adverbs describe or add information about other words in phrases and clauses.

An **adjective** modifies the meaning of a noun or a pronoun. In other words, adjectives provide additional information so that the reader or listener gets a clearer, more complete idea of what the noun or pronoun signifies. Here are some examples that modify nouns:

complete coverage *heavy* load *blue* uniform *inexact* measurement

These adjectives modify pronouns:

lucky you I am *late* they look *angry* he was *brilliant*

Most adjectives that modify pronouns follow linking verbs, as in "Each type of error is avoidable."

Adverbs modify the meaning of verbs, verbals, adjectives, or other adverbs, as in the following examples (all adverbs are in italics):

adverbs that modify verbs— run *swiftly* write *carefully* jumped *gracefully*

adverbs that modify verbals— to aim *correctly* to advance *slowly*

adverbs that modify adjectives—*so* tall *very* slow extra *careful*

adverbs that modify adverbs— *very well* *too carefully* *too highly* placed

Types of adjectives

Several types of purposes are served by adjectives:

adjectives that **describe**— *red* earth *large* fish *knotted* rope

adjectives that **limit**— *two* collisions *occasional* crashes *my* problem

adjectives that **emphasize**— the *very* one *perfect* success *complete* idiot

English has several word sources for adjectives. Some words function only as adjectives:

tall splendid large empty strong rich

Others can also function as nouns:

cold rough computer wool car football.

Still others spend part of their time as adverbs:

well late early hard near daily

Some adjectives are taken from the participle form of verbs:

washing machine parking meter whistling wind running water

Other adjectives take hyphenated forms:

last-minute orders *up-to-date* equipment *well-built* truck *10-year* employee

Articles as adjectives

Many grammarians classify the articles *a*, *an*, and *the* as adjectives because these useful little words limit the meaning of nouns. The definite article *the* specifies a particular, identifiable noun as in *the causeway* or *the apple*, while the indefinite articles *a* and *an* refer to nouns in nonspecific ways, as in *a causeway* or *an apple*.

Types of adverbs

Most adverbs are formed by adding *ly* to adjectives or by dropping or changing the final letter and then adding *ly*:

Adjective	Adverb
sad	sadly
efficient	efficiently
cold	coldly
slow	slowly
quickly	quickly
due	duly
empty	emptily
busy	busily

But some adverbs do not end in *ly*:

very soon too much often well

Some adverbs are formed as phrases, not as single words:

They displayed the waveform *on an oscilloscope.* [The preposition phrase *on an oscilloscope* is used as an adverb to modify *displayed.*]

I'm writing this book *to showcase effective grammar.* [The infinitive phrase *to showcase* is used as an adverb to modify *writing.*]

Occasionally, the adverb appears as a clause that contains a subject and a verb:

We reported the incident *after we returned to the office.* [*After we returned to the office* modifies *reported* by indicating when the report happened.]

Comparative forms of adjectives and adverbs

Both adjectives and adverbs can be used to compare items. Grammarians use the terms *positive, comparative, and superlative* to denote the three degrees of comparison. Table 6 lists adjectives and adverbs in these three degrees.

> **Note:**
> Some grammarians use the term *simple* instead of *positive*.

Table 6 Comparative Forms

	Positive	Comparative	Superlative
Adjectives	hard	harder	hardest
	tall	taller	tallest
	rich	richer	richest
	good	better	best
	little	less	least

	some	more	most
	acceptable	more acceptable	most acceptable
	daunting	more daunting	most daunting
Adverbs	fast	faster	fastest
	well	better	best
	badly	worse	worst
	slowly	more slowly	most slowly
	aggressively	more aggressively	most aggressively
	smoothly	less smoothly	least smoothly

Note:

Most adverbs use more and most or less and least to create their comparative forms.

Usage hints

The following advice is designed to help you use adjectives and adverbs effectively.

1. Use *a* and *an* properly

Use *a* when the following word starts with a consonant sound:

a bracket *a* design *a* software program *a* CBC podcast

Use *an* when the following word starts with a vowel sound:

an amateur *an* honour *an* SOS signal [The *S* here is sounded as "es."]

Until recently, *an* was used with unaccented syllables beginning with *h*, as in "*an* history of Canada"; however, current usage is switching to *a* history or *a* hexagonal design.

2. Be careful with comparative adjectives and adverbs

Most one-syllable adjectives use *-er* and *-est* endings for comparatives:

quick quicker quickest

Some two-syllable adjectives take *-er* and *-est*; others use *more* and *most*:

windy windier windiest

grateful more grateful most grateful

Some two-syllable adjectives take *-er* and *-est*; others use *more* and *most*:

windy windier windiest

grateful more grateful most grateful

Adverbs most often use *more* and *most* or *less* and *least* to form comparatives:

clearly more clearly most clearly

Three-syllable adjectives and adverbs almost always use *more* and *most* or *less* and *least* to form comparatives:

honourable	more honourable	most honourable
cheerfully	more cheerfully	most cheerfully

Avoid double comparisons in which both *-er* and *more* (or *-est* and *most*) are used:

The ~~more~~ windier city   the ~~most~~ richest dessert on the table

3. Avoid double negative adverbs

Truly effective negative adverbs such as *neither*, *nor*, *no*, *not*, *scarcely*, and *seldom* don't need help in conveying negation. So avoid statements such as the following (especially if you want to seem educated):

I *don't* want *no* more materials delivered to the Spadina worksite. [Write "I *don't* want *any* more material ..." or "I want *no* more materials ..."]

They *couldn't hardly* cross the road because of the heavy traffic. [Write "They *could hardly* cross ..." or "They *couldn't* cross ..."]

Raphael scarcely paid attention to neither one of the inspectors. [Write "Raphael *scarcely* paid attention to the inspectors" or "Raphael paid attention to *neither* of the inspectors."]

4. Be aware of the pitfalls with *well*, *good*, *bad*, *badly*, *real*, and *really*

Well can function as an adjective or as an adverb. And, in conversation, many people don't pay attention to the distinctions between *good* and *well* and between *bad* and *badly*. Here are some examples of ungrammatical usage:

You did really *good* on the Martin Brothers project. [Correct usage: "You did really *well* ..."]

Marta performed *bad* on her job interview. [Correct usage: "Marta performed *well* ..."]

The lobster tasted *badly* because of the rancid garlic butter. [Correct usage: "The lobster tasted *bad* ..."]

He uses spreadsheets *good* and accounting software *not bad.* [Correct usage: "He uses spreadsheets *well* and accounting software *not badly.*"]

He's been feeling *good* since his kidney transplant. [Correct usage: "He's been feeling *well* ..."]

You look *well* in that black dress. [Correct usage: "You look *good* ..."]

Real and *really* are often used incorrectly in casual conversation:

He has a *real* good attitude. [Correct usage requires an adverb: "He has a *really* good attitude."]

Remember that *real* is an adjective and *really* is an adverb.

5. Do not pile up noun modifiers

Nouns can function as adjectives in such phrases as *horse farm* or *atom bomb*, but when noun modifiers pile up, they can obscure meaning. For example, how long would it take a reader to decipher this phrase: *duplicate cross-contaminant groundwater well samples*? Four nouns pile up in front of *samples*, and two of those nouns used as adjectives are compound nouns. The poor reader looks at such a phrase and thinks, "come on, give me a break!" Here are two more examples taken from actual workplace documents:

The *2007 hydrology survey access* information contained no surprises. [In this sentence, four nouns all help modify *information.* Four!]

The board created a *downtown community access and decision* structure to deal with the burgeoning rabbit population. [This sentence appeared in a newspaper account of a City Council's creation of a committee. Yes—*committee* was what they meant by all those words.]

6. Know when to use commas to separate adjectives

When more than one adjective precedes a noun, commas can be very useful in signalling the writer's exact meaning:

They discovered toxic, non-recoverable traces of pharmaceuticals in the water supply. [*Toxic* and *non-recoverable* independently modify *traces*; *toxic* does not modify *non-recoverable*.]

They discovered toxic pharmaceutical elements in the water supply. [Here, *pharmaceutical* modifies *elements* and *toxic* modifies *pharmaceutical*.]

Hint:

If you could change the order of the adjectives with no real change in meaning, these adjectives are acting independently and require commas:

Several *industrious, motivated, high-achieving* students received scholarships.

But if reversing the order would change the meaning, you shouldn't use a comma:

Our *extended equipment* warranty does not cover damage caused by negligence.

Practice session—identify adjectives and adverbs

Underline the adjectives and circle the adverbs in the following sentences. You will find answers for these exercises on pages 119 and 120.

1. Two dedicated teams desperately want to win the Vanier Cup, which is awarded annually to the champion of CIS football.
2. Biodiesel has a higher cetane level than petroleum diesel.
3. The Canadian cement industry claims that it has steadily improved energy efficiency in its manufacturing processes.
4. Improved indoor air quality probably reduces the number of sick days claimed by workers.
5. Despite urgent requests from the local health authorities, the mine repeatedly refused to install modern water treatment methods.
6. He has done well with the aging equipment at his disposal.
7. The Earth Home system provides fully integrated, self-sufficient buildings.
8. Compressed earth bricks are environmentally sustainable because they require comparatively little energy to produce and very little energy to transport.

9. The sales manager will work closely with the plant manager to plan just-in-time deliveries.

10. Someone has inadvertently left the specimen case on the loading dock.

Improvement exercises—adjectives and adverbs

In items 1 through 10, repair errors in the use of adjectives and adverbs. If a sentence contains no errors, write ***C*** in the space provided. You will find answers for these exercises on page 120.

1. Don't think bad of me because I tried to improve our working conditions.

2. According to Bobbie Cruz in the IT department, every computer, hard drive eventual fails.

3. They advised us to move slow and careful when working with those volatile chemicals.

4. We must conserve depleted scarce non-renewable resources or face the consequences.

5. Reported, rollover rates for vans and SUVs are considerably higher than for sedans.

6. He was clearly the strongest of the two of them.

7. The maple syrup harvest is way more early than usual this year, perhaps because of global warming.

8. The researchers made a honest mistake when they miscalculated the size of the honey bee population in Ontario.

9. Global cooling sure is more likely than global warming in the next 50 years.

10. Windows Vista arrived on the market with more glitches than I was prepared to accept.

Each of the sentences 11 through 20 contains a highlighted adjective or adverb. In the space below the sentence,

a. identify whether the word is being used as an adjective or as an adverb, and

b. write a sentence that uses that word as the other type of modifier. (If the highlighted word is an adjective, use it as an adverb; if the word is an adverb, use it as an adjective in the sentence that you write.) If necessary, consult a dictionary's list of meanings for the word. See page 120 for sample answers.

11. In March 2009, Northern Landscaping discovered a <u>near</u> collapse of the Arduin Reservoir dike.

12. He is usually <u>late</u> for meetings that require him to present a project update.

13. Please report <u>daily</u> about the progress of labour contract negotiations.

14. Three times this season he has been reprimanded for playing <u>fast</u> and loose with the fishery regulations.

15. She responded to the dissident workers' grievances very well, don't you agree?
16. Please submit the monthly production figures by the second day of the following month.
17. Do you think that our boss is a well woman?
18. The early indications are that the real estate bubble is about to burst.
19. Just how hard did you try to make the exam questions difficult?
20. *The Globe and Mail*'s weekly "Career" section has recently focused on the changing demographics of the Canadian work force.

Document critique—adjectives and adverbs

The following document may contain errors in grammar, capitalization, punctuation, or spelling. Underline all errors and label them using standard proofreading marks. (See the Correction Symbols list on page 144.) You will find answers for this exercise on page 121.

Christopher Metals

MEMO

To: Shift Foremen

From: Augusta Brown, General Manager

Date: May 22, 2009

Subject: Reducing Toxic Emmisions At Our Plant

As you have might have heard, we received a D^- from Environment Canada following their December site visit. Apparently we have 6 months to clean up are act—the inspecters will be here again in November. If are sulphur and hydrochloric acid readings have not been reduced by 50 percent, they'll move to close us down.

To be honest, I don't think we've bin doing that bad. For example, are culvert mill and are lead coating facility doesn't have no more free-flowing smoke stacks—the emission capture equipment diverts the gases to the holding chamber wear the gases is chilled and the toxic elements are filtered out. But that's not good enough for the government, I guess.

I'm attaching some literature on equipment thats supposed to reduce sulphuric and hydrochloric acid emmisions. Would you please read the descriptions and look at the diagrams of the equipment to see whether we could fit such devices into are existing setups? I mean, you know you're areas inside out, so youll know if the equipment will work in are plant.

Pay particuler atention to the rugged low-cost low-maintenance air pollution control equipment sold by Herring Engineering. How much of it would we need? Can we cut into are ductwork to install it? The price looks right, so look at the Herring stuff real carefully

Lets discuss this next Tuesday in are regular production meeting.

Conjunctions, Prepositions, and Interjections

Three kinds of helping words are discussed in this section: conjunctions, prepositions, and interjections help complete the patterns of meaning conveyed by nouns, pronouns, and verbs.

Conjunctions

These connecting words reveal three kinds of relationships—*coordinating*, *correlative*, or *subordinating*.

> Ricky Ruiz *and* Elma deLoos had their laptops stolen *because* the computers were not properly stored. Perhaps that's why *neither* Ricky *nor* Elma has yet claimed for insurance.

Coordinating conjunctions such as *and* join equivalent sets of nouns, pronouns, verbs, adjectives, adverbs, prepositions, phrases, and clauses. (Pages 45 and 46 define phrases and clauses.)

Correlative conjunctions such as *neither* and *nor* work in pairs to establish a connection.

Subordinating conjunctions such as *because* introduce dependent (a.k.a. subordinate) clauses.

Table 7 Conjunctions and Their Uses

Coordinating	Relationship They Express
and,	addition
but, yet	contrast
for	cause or reason
so	effect or result
or	choice
nor	negative choice

Correlative		Relationship They Express
both ... and	not only ... but	addition
either ... or	whether ... or	contrast
neither ... nor		negative choice

Subordinating	Relationship They Express
because	cause or reason
in order that, so, so that, that	effect or result
after, before, since, until, once, when, whenever, while	time
if, even if, provided that, unless	condition
although, even though, whereas	contrast
where, wherever	location

than, whether	choice

Note:

See the "Sentence Grammar" section for examples of appropriate and inappropriate coordination and subordination.

Conjunctive adverbs create logical connections within sentences. At the same time, they modify verbs, adjectives, or other adverbs within the sentence. Table 8 lists conjunctive adverbs.

Table 8 Conjunctive Adverbs

Words	Relationship They Express
also, furthermore, moreover, besides	addition
however, still, nevertheless, conversely, nonetheless, instead, otherwise	contrast
similarly, likewise	comparison
therefore, thus, consequently, accordingly, hence, then	result or summary
next, then, subsequently, meanwhile, finally	time
indeed, certainly	emphasis

Conjunctive adverbs such as *however* can effectively operate at the beginning, in the middle, or at the end of sentences:

However, people with the HIV virus find their weakened immune system significantly improved by this treatment regimen.

People with the HIV virus, *however*, find their weakened immune system significantly improved by this treatment regimen.

People with the HIV virus find their weakened immune system significantly improved by this treatment regimen, *however*.

Prepositions

Most prepositions convey relationships in time or space, between nouns or pronouns and other words. Typical examples of prepositions appear in the following phrases:

in a year   *in* a few minutes   *in* the morning   *at* noon *on* Friday

in town *in* Nova Scotia   at the ballpark   *on* the road   *on* Main Street

Here are some other prepositions:

about, across, after, along, among, against, apart from, as, behind, beneath, between, beyond, down, from, in front of, into, on, onto, past, round, through, toward, until, up, with, without

Prepositional phrases essentially function as adjectives or as adverbs:

I'll return his call *in* a few minutes. [adverb—indicates **when**]

The picture *above* your desk [adjective—indicates **where**]

reminds me *of* my lake cottage. [adverb—indicates **what**]

Preposition usage hints

Conventions for using prepositions and prepositional phrases have evolved through centuries of usage. One of those "rules," which prohibits placing prepositions at the end of sentences, has been relaxed because some expressions are downright silly when that rule is strictly enforced. Winston Churchill dramatized that silliness in his famous lampoon line "This is something up with which I will not put!" Churchill knew that "This is something I won't put up with" or "I won't put up with this" are more natural and less pedantic. Still, we should try to avoid ending sentences with prepositions, especially if that final preposition is not really needed, as in

Where was Roger going ~~to~~? Or, What are you thinking ~~of~~?

Superfluous prepositions at the beginning or in the middle of sentences can safely be deleted:

He fell off ~~of~~ the Ferris wheel at the CNE. ~~Like~~, he couldn't help ~~from~~ showing off.

Many distinctive English idiomatic expressions[3] include prepositions:

above the fray	*under* the weather	*behind* the 8-ball	*down* to the wire
by all means	*by* a country mile	*at* the drop of a hat	*against* the grain

Writers may confuse these pairs of prepositions and phrases:

- ***among*** and ***between***—*between* refers to two (the road *between* the poplar bluff and the creek), while *among* refers to three or more (share the research load *among* George, Paula, and Syd)
- ***in*** and ***into***—in refers to static positions (the dump truck *in* the yard); *into* refers to movement toward a position (despite the tight opening, we crawled *into* the cave)
- ***as if*** and ***like***—*as if* comes before a clause (it's not *as if* he has seniority)
- ***beside*** and ***besides***—*beside* means "next to" (place the barrels *beside* the loading dock); *besides* means "in addition to" (we have other alternatives *besides* declaring bankruptcy)
- ***different than*** and ***different from***—both these phrases are widely used, especially in speech and in informal writing, but *different from* is generally preferred in formal writing
- ***differ from*** and ***differ with***—*differ from* means to be unlike something (the 2008 Acura MDX *differs from* the previous model in its styling and its engine); *differ with* means to disagree (I *differ with* his interpretation of the landslide's causes)
- ***like*** and ***such as***—*like* refers to the similarities between two or more compared items (the Honda Accord, *like* the Acura TL, uses a V-Tec 6-cylinder engine); *such as* often replaces *for example* in order to achieve smoother phrasing (natural disasters *such as* hurricanes are becoming more frequent)

Interjections

These words exclaim; they present emotion:

---

[3] Wayne Magnuson's comprehensive and entertaining *English Idioms: Sayings and Slang* (Calgary: Prairie House Books, 2001) is a very useful compendium of English idioms, with special emphasis on North American sayings.

Oh, I don't think I want to buy that car. *Oh, no!* We have another broken hex bolt.

Bravo! We needed to win that job, and you've secured the contract with your well-written proposal.

When interjections appear on their own, typically they're followed by an exclamation point. When they appear in a sentence, usually they're set off by one or more commas. Note, however, that interjections don't appear very often in technical writing.

Practice session 1—identify prepositions

Underline the prepositions in the following sentences. Then, circle the complete phrase in which each preposition appears. You will find answers for these exercises on page 121.

1. The Vanier Cup is awarded annually to the champion of CIS football.
2. Improved indoor air quality probably reduces the number of sick days claimed by workers.
3. Despite urgent requests from the local health authorities, the mine repeatedly refused to install modern water treatment methods.
4. The news media are very interested in the global warming issue.
5. Someone has inadvertently left the specimen case on the loading dock.

Practice session 2—identify conjunctions

Using the supplied form, list and name all the conjunctions in the following passage and describe the relationship each one expresses. An example is provided. (See Table 7, page 38.) You will find answers for these exercises on page 122.

> New research and increased understanding of natural cycles suggest that longer autumns may be diminishing the ability of northern forests to fight global warming. Trees absorb carbon dioxide from the atmosphere as they grow, but they release CO_2 when they burn or decompose. Spring warming trends have increased tree growth, so more CO_2 is being absorbed than "normal." However, autumns have been warming at an even greater rate than springs, so more CO_2 is being released than normal because soil microbes are more active in the warmer weather and their activity releases carbon dioxide.

Conjunction	Type of Conjunction	Relationship Expressed
and	coordinating	Connects the compound subject "research and understanding"

Practice session 3—identify prepositional functions

Using the supplied form, identify the prepositional phrases and the nature of their modifying functions in the following passage. An example is provided in the form. You will find answers for these exercises on page 122

> The day after a burrowing owl captured media attention in Ottawa, a 6,000-name petition was tabled yesterday in the House of Commons. The petition supports plans for a national park in the South Okanagan. The petition was submitted by B.C. Southern MP Alex Atamanenko, prior to a meeting with representatives of the National Park Network. Atamanenko said that he and his staff have worked closely with Parks Canada officials to facilitate a dialogue between groups that hold opposite points of view regarding the grasslands park proposal.

Prepositional Phrase	Type of Modifying Phrase	Word or Phrase Modified
in Ottawa	adverb	Says where the owl "captured media attention"

Improvement exercises—conjunctions, prepositions, and interjections

In items 1 through 10, repair errors in the use of conjunctions, prepositions, and interjections. (You may also have to fix punctuation or spelling errors.) If a sentence contains no errors, write ***C*** in the space that follows the sentence. You will find answers for these exercises on pages 122 and 123.

1. Where is the new computer projection system at?
2. Clearly, we have to improve the communication between the production team, the sales people and the shipping people.
3. Either the transmission bolt threads are stripping when the engine torque increases because they're made of too soft steel; the engine might also be too powerful for this type of transmission.
4. Michelle damaged the portable oscilloscope when she tried to jam it in a metal container that wasn't large enough.
5. Please remember to not place the GPS receiver besides equipment that emits a strong magnetic field.
6. The maple syrup harvest involves like way more work than most people realize.
7. Way to go, that set of schematics represents the best work you've done for our firm.
8. Big hybrid SUVs, like the Toyota Highlander, the Chevrolet Tahoe, and the GMC Yukon, are now getting up to 30 miles per Imperial gallon.
9. Neither the diehard Ottawa football fans or the corporate community believe that a CFL football team can flourish in Ottawa.
10. A large body of evidence supports the contention that global warming results primarily from sun spot activity, not from fossil fuel emissions, however.

Document critique—prepositions and conjunctions

The following document may contain errors in grammar, capitalization, punctuation, or spelling. Underline all errors and label them using standard proofreading marks. (See the Correction Symbols list on page 144.) You will find answers for these exercises on page 123.

Christopher Metals

MEMO

To: Shift Foremen
From: Augusta Brown, General Manager
Date: June 2, 2009
Subject: Reducing Toxic Emmisions At Our Plant

Thank you for you're useful coments in yesterdays meeting. I know we differ with each other on some points, but overall are experience point to ward either of three alternative.

Alternative 1 – upgrade are the emission capture equipment. I thought that are existing equipment is state of the-art, however, as Jim reported, its already showing signs of aging and needs new filtering and storage addons. This alternative is probably the least expensive.

Alternative 2 – shut down the galvanizing facility. This section of the plant requires the most expensive work, still it generates 65 percent of are income. So I don't want to shut it down.

Alternative 3 – shut down the culvert mill, sell its equipment, and use the sale funds to expand and modernize are galvanizing facility. Not only would this strategy get rid of the culvert mill, that is bearly coverring operating costs and it would allow us to expand production of galvanized pipe, that is in high demand. Beside these benefits, Jeanine pointed out that new ways of capturing and storing galvanizing by-products have been develop.

Their seems to be more research to do, as we look in this issue. Lets meet again on Monday.

SENTENCE GRAMMAR

The previous chapter, "Basic Grammar," focuses on the role of individual words and small groups of words in conveying meaning. This chapter studies the word patterns that effectively describe, explain, and persuade. Make no mistake about it—good technical writing depends on these patterns as much as it depends on the words themselves.

As in the previous chapter, our discussion of sentence grammar is descriptive, not prescriptive, as this book passes on the wisdom and techniques learned by technical writers. Countless writers have learned the hard way that too many long, complex sentences make readers weary, that dangling modifiers confuse readers, or that lack of parallelism leaves readers scratching their heads. This chapter covers those three topics and more, but first let's look at main structural principles that make English sentences work.

Sentence Building Blocks

At the heart of the English sentence are **subjects** and **predicates**. If a group of words contains both of those elements, we call the grouping a **clause**. A one-word subject mated with a one-word predicate (such as *he runs* or *commerce rules*.) qualifies as a clause, but such two-word sentences are rather rare, especially in technical writing.

More often, subjects and predicates also contain adjectives, adverbs, conjunctions and/or phrases that contribute to more detailed information and to more complex thought.

Grammar at Work: clear communication saves money and reputations

An event planner sent information about a national conference to all the members, vendors and speakers attending the conference, but she wrote the wrong start date of the conference in her message. As a result, her company had to pay nearly $20,000 in airline change fees and other penalties. Such mistakes can cost time and money and can erode a company's credibility.

> "In addition to credibility issues, the wrong dollar amount in a contract or letter of agreement may force one party to honor the wrong price.... The wrong address or phone number can delay or prevent contact or correspondence. These errors may be considered 'minor' but they can potentially be costly and embarrassing." (Ronnie Moore, in *Why Did I Say That? Communicating to Keep Your Credibility, Your Cool, and Your Cash!*)

—Based on "Sloppy Writing at Work Has Dire Consequences," a report by ***Rachel Zupek****, Careerbuilder.Com Writer, Posted March 3, 2008 at http://jobs.aol.com/article/_a/sloppy-writing-at-work-has-dire/20071115155809990001*

Subjects

Grammarians distinguish between a **simple subject** (a key word or group of words that acts, or describes a state of being, or is acted upon) and a **complete subject** (the simple subject and its modifiers). In each of the following sentence, the complete subject is italicized and the simple subject is underlined.

Every scientist at the research station has resigned.

Our language's amazing versatility is reflected in the variety of forms that subjects can take:

Single noun: *Music* soothes the soul.

Single pronoun: *She* repaired the fuel line.

Compound subject: The white *van* and the *Hummer* crashed head on.

Noun phrase: *Corporate responsibility* has tilted toward satisfying the shareholders.

Gerund (a present participle form of a verb): *Running* gives me time by myself.

Gerund phrase (a gerund and other words combine to act as a subject): *Running a drag line* requires specialized training.

Note:

Commands such as *Sweep the deck* or *Submit the proposal by Monday morning* usually don't include the subject that is understood to be present, *you*. In rare cases, the subject is inserted in a command sentence: *Ladies and gentlemen*, *please rise and remove your hats for the singing of the national anthem.*

Predicates

In a sentence, the predicate contains the verb, which expresses what the subject is experiencing or doing, or what is being done to the subject. A **simple predicate** contains only the verb:

Our team leader is *ill.* The train *stopped.* Surprisingly, the Canadian dollar *rose.*

A **complete predicate** contains the simple predicate and one or more modifiers:

Our team leader is very ill. The train stopped abruptly at the station.

A **compound predicate** contains two or more verbs:

The coach *waved* her arms and *whistled* in a vain attempt to get the point guard's attention.

Our latest ice storm *damaged* trees and *downed* power lines.

Clauses

The above examples of sentences show **independent clauses**, patterns of thought that stand alone as complete sentences. Although **dependent clauses** also contain a subject and a predicate, they rely on being attached to an independent clause to complete their meaning. In the following sentence, the dependent clause (in italics) comes first, followed by the independent clause that could have operated on its own:

After he left university, Raoul joined an oil exploration firm.

In the next example, the independent (or main) clause comes first, followed by the dependent (or subordinate) clause:

Sarah won a Kohlman bursary because her grades were well above the norm.

Dependent clauses that perform an adverbial function usually start with a subordinate conjunction such as *because*, *until*, *after*, or *when*. For a more complete list of subordinate conjunctions, see Table 7, page 38.

Types of Sentences

Grammarians recognize four types of sentences:

- Simple sentences contain a single independent clause;
- Compound sentences have two independent clauses;
- Complex sentences contain one independent and one or more dependent clauses; and
- Compound-complex sentences feature two or more independent and one or more dependent clauses.

Simple sentences

Good technical writing contains a high percentage of straightforward sentences. One study[4] found that readable prose consists of a minimum of 75.5 percent simple sentences, either in **subject-verb-object** (SVO) order or in **subject-verb-complement** (SVC) order.

SVO sentences are easy to follow, especially if the subject (underlined below) appears at the very beginning of the sentence, followed immediately by the verb (italicized below):

> Regina *will host* the 2010 Grey Cup game.
>
> He *won* the first match with a devastating serve.

Both the above examples feature simple subjects, but sentences with more complete subjects are also quite easy to understand:

> Actually, Regina, the home of the Saskatchewan Roughriders, *will host* the 2010 Grey Cup game.
>
> Later that afternoon, he easily *won* his second match with a finesse game, proving his adaptability.

These embellished sentences now begin with a transitional word or phrase that helps maintain a smooth flow of ideas within a paragraph. The first sentence also inserts a noun phrase (*the home of the Saskatchewan Roughriders*) that adds some texture to the sentence. The second sentence adds an adverb (*easily*) and a verbal phrase (*proving his adaptability*). Still, these augmented SVO sentences remain easy to follow.

Most **SVC sentences** are also easy to read—the complementary bits follow the core subject and verb. The complement could say more about the subject, or more about the verb, or both.

- An adjective or phrase could function as an adjective, following a linking verb:

  Our boss is *unhappy*. [*Unhappy* modifies *boss*.]

  Martine is *under the weather*. [The prepositional phrase *under the weather* modifies *Martine*.)

- Adverbs or adverbial phrases might follow an active verb:

  Denman works *hard at his chosen profession*, *for six 12–hour days a week*. [The adverb *hard* modifies *works*. The prepositional phrase *at his chosen profession* further modifies *works*. The phrase *for six 12-hour days a week says* **when** Denman works.]

---

[4] Dr. Francis Christensen, an English professor at the University of Southern California, reported his findings in "Notes toward a New Rhetoric," in *College English*, October 1963, pp. 7–18. His data and conclusions remain relevant today.

- A string of complementary phrases might combine adverbial and adjectival functions:

  On Sundays, Denman remains *happily at home with his collection of exotic birds*. [The adverb *happily* modifies *remains*. Two adverbial prepositional phrases complement the subject-verb core of this sentence: *at home* and *with his collection*. Both say **where** Denman *remains*. The adjectival phrase *of exotic birds* modifies *collection*. This **cumulative** building of meaning works well because it keeps the core meaning at the beginning of the sentence, and it rhythmically adds information.]

The previous example uses a short prepositional phrase, *on Sundays*, to lead into the sentence. Such transitions perform a useful function and they don't impede the reader's understanding of the sentence's meaning.

The following paragraph illustrates SVO and SVC sentences at work in a technical document:

> On Lake Superior, Environment Canada annually samples water at 94 stations. This sampling requires a sturdy vessel to navigate the lake and gain access to remote spots along the shore. In addition, 25 surveillance cruises have gathered data on Lake Superior between 1976 and 2006. During these cruises, scientists sampled for nutrients, major ions, organic contaminants, and biological and physical trends. The results have been startling.

The above passage packs a considerable amount of information into its 66 words. However, it's readable because all five sentences use an SVO or SVC pattern, and the sentences are linked with transitions such as "this sampling," "in addition," and "during these cruises." To further aid readability, the average sentence is only 13.2 words long, the average number of words before the subject is only 2.0, and the verbs immediately follow those subjects.

The Good News:

You don't have to be a great stylist to be an effective technical writer; you just have to write lots of clear SVO and SVC sentences, linked by transitional words and short transitional phrases.

Compound sentences

At some point in your writing career, you've probably been told that you need to write a variety of sentence types, including compound and complex sentences. That advice doesn't necessarily hold for technical writing, but sometimes your thoughts will require longer, more involved sentence structures such as compound sentences. Here are two compound sentences, one with two independent clauses and one with three:

> Servers must operate continuously and reliably, so <u>efficient, trustworthy fans must be used for ventilation</u>.
>
> <u>Some popular operating systems for servers are derived from the UNIX operating system</u> and <u>Unix has been a logical choice for many servers</u>, but *the market share of the Windows Server product line has been growing steadily.*

The 15-word sentence doesn't present a reading challenge, but the 37-word sentence demands close attention, perhaps even a second reading. Writers of such sentences need to plan them carefully to ensure that the thought pattern develops logically.

Complex sentences

Essayists who espouse complex philosophies often need complex modes of expression—the complex sentence, with its single independent clause and one or more dependent clauses, can serve that need. However, sentences such as the following are relatively rare in technical writing:

> Although electrical signals propagate along a cable very quickly, <u>they weaken as they travel</u>, so that electrical interference from neaerby devices such as fluorescent lights can scramble the signal.

The independent clause "they weaken as they travel" is at the heart of the above complex sentence.

Now, are you ready for an even more complicated collection of words?

Compound-complex sentences

This type of sentence contains two or more independent clauses and one or more dependent clauses, as in:

> A single shared cable can serve as the basis for a complete Ethernet network and several machines can be connected to the network, although there are practical limits on the size of the Ethernet network in this case, and the length of the shared cable must be considered.

So here we have 4 clauses and 48 words in one sentence. The sentence requires your full commitment, doesn't it? Would you feel better if the sentence were broken up? Do you find the following arrangement easier to read?

> A single shared cable can serve as the basis for a complete Ethernet network and several machines can be connected to the network. However, this case imposes practical limits on the size of the Ethernet network. In particular, the length of the shared cable must be considered.

Now the four clauses occupy three sentences and 47 words. Flow has been maintained by using transitions.

Sentence Errors and Challenges

Relatively short, uncomplicated sentences help readers navigate technical and business documents, but these sentences can still be sabotaged by the following kinds of mistakes.

Packed sentences

Technical writers may be tempted to overfill sentences with descriptive or analytical detail:

> The Nokian WR all-season tire's features provide important safety benefits: a high dispersion silica compound provides excellent grip, longer mileage, and better rolling resistance while the arrowhead tread pattern effectively expels water, snow, and slush, and the Haka siping creates "rubber studs" in the tread that result in better grip and high stability during high speed driving.

I defy you to grasp that 57-word statement the first time you read it. Perhaps a bulleted list would make the sentence easier to swallow.

> The Nokian WR all-season tire features important safety benefits:
>
> - a high dispersion silica compound provides excellent grip, longer mileage, and better rolling resistance;
> - the arrowhead tread pattern effectively expels water, snow, and slush; and

- the Haka siping creates "rubber studs" in the tread that result in better grip and high stability during high speed driving.

Words before the subject

Subjects can be hard to find in sentences that open with long phrases.

> In a city where careers are made and broken in the time it takes to grill a hot dog, *Pink's* is one of Hollywood's few constants—a place where families, struggling actors and the show-biz elite temporarily unite in the common pursuit of hot dogs.

The above statement trots out 19 words before getting to the subject of the sentence, *Pink's*. Elaborate openings might be fine for popular journalism—the above sentence appeared in an MSN story about America's best hot dog joints—but readers of technical writing don't expect to be entertained; they want plain English and they want sentences to place the subject at or near the beginning, as in this sentence:

> The Nokian WR all-season tire features important safety benefits:
>
> - a high dispersion silica compound provides excellent grip, longer mileage, and better rolling resistance;
> - the arrowhead tread pattern effectively expels water, snow, and slush; and
> - the Haka siping creates "rubber studs" in the tread that result in better grip and high stability during high speed driving.

Compare the above list to this mess, which presents 52 words before naming the subject:

> Because of excellent grip, longer mileage, and better rolling resistance resulting from a high dispersion compound, together with an the arrowhead tread pattern that effectively expels water, snow, and slush and Haka siping that creates "rubber studs" in the tread that result in better grip and high stability during high speed driving, *the Nokian WR all-season tire* rolls out important safety benefits.

Here's a good rule of thumb: The sentences in a given passage should average no more than three words before naming the sentence's subject. In other words, get to the point as soon as possible.

Words between the subject and the verb

Again, the rule of thumb uses the number *3*. On average, no more than three words should come between the subject and the verb in clauses. If you use more, you increase reading difficulty. When you make material hard to read, you lose your readers.

> Employment portfolios, which might contain supporting documents, work samples, lists of accomplishments, records of community service, photo documentation, certainly a targeted résumé, and maybe even video evidence, can be very useful in job searches.

After ploughing through the 27 words that separate the subject and the verb, the reader might well forget what the subject of the sentence is. A better approach would reunite the subject and verb at the beginning of a two-sentence statement:

> Employment portfolios can be very useful in job searches. These portfolios might contain supporting documents, work samples, lists of accomplishments, records of community service, photo documentation, a targeted résumé, and maybe even video evidence.

And why not break the second statement into a bulleted list?

> Employment portfolios can be very useful in job searches. These portfolios might contain

- supporting documents and photo documentation,
- work samples,
- lists of accomplishments,
- records of community service,
- a targeted résumé, and
- video evidence.

Sentence fragments

A phrase or a dependent clause that is presented as if it were a sentence is called a **sentence fragment**. Editors pounce on these fragments because complete thoughts are not provided. Still, fragments have their place in, for example, advertising messages:

> Come ski at the Rocking Donkey Resort. The best powder in B.C. No lift lines. And best bang for your buck!

The above example would lack punch if the ad read, "We have the best powder in B.C. along with virtually no lift lines. Therefore, we offer excellent value." No, that approach doesn't work in ad copy. However, completed thought is exactly what's needed in workplace writing, which needs to describe and explain, not entertain. Here's an example of the problem with sentence fragments in workplace writing:

> Pollutants can enter a watercourse through a particular entry such as a drainpipe*. Flowing directly into a stream or lake. Point source pollution.* That's its name, and its main sources are sewage treatment plants and industrial waste discharges. *Including toxic chemicals and human waste.*

In the above paragraph, the three fragments are italicized. Each lacks an element that would clarify what the writer intended to convey. It's hard to tell if that first fragment is meant to say more about the preceding noun, *drainpipe*, or whether *flowing directly into a stream or lake* is meant to be tied to the phrase, *point source pollution*. The third fragment might be intended to modify *discharges*, or it might be a modifying phrase for a statement that didn't make it to the page. The following edited version adds a clause that can show the relevance of *toxic chemicals and human waste*.

> Pollutants can enter a watercourse through a particular entry such as a drainpipe that flows directly into a stream or lake. Known as point source pollution, its main sources are sewage treatment plants and industrial waste discharges. Point source pollutants include toxic chemicals and human waste.

Let's quickly review. Phrases and dependent clauses aren't particularly useful by themselves. For example, *because atmospheric pollution comes from many diffuse sources* is incomplete, even though it contains a subject and a predicate. The conjunction *because* has hinted at something that would complete the thought, but we don't know what that something is. We could guess that the intended complete sentence would read *Because atmospheric pollution comes from many diffuse sources, it is difficult to control*. However, guesswork isn't good enough for technical and business documents.

Run-on sentences

Complete sentences need to be separated by periods, semicolons, or colons. These punctuation methods receive full treatment in the "Punctuation Exercises" section of this workbook, so here we focus on what's wrong with **comma splices** and **fused sentences**.

> Come ski at the Rocking Donkey Resort, it has the best powder in B.C. and no lift lines.

I've underlined the sequence of words where the two independent clauses are ineffectively *spliced* together with a comma. The first clause ends at *resort*. The second clause begins at *it*. That comma doesn't help us at all. Now here are four possible interpretations of the writer's intention:

1. Come ski at the Rocking Donkey Resort *because it* has the best powder in B.C. and no lift lines.
2. Come ski at the Rocking Donkey *Resort, which* has the best powder in B.C. and no lift lines.
3. Come ski at the Rocking Donkey *Resort. It* has the best powder in B.C. and no lift lines.
4. Come ski at the Rocking Donkey *Resort; it* has the best powder in B.C. and no lift lines.

These four versions have slightly different meanings. Version 1 emphasizes the cause-effect relationship. Version 2, like Version 1, also turns the second clause into a dependent clause, but now the reasons for going to the Resort are less important than going there. The second clause is no longer essential to the sentence. Versions 3 and 4 separate the two clauses, but the period in Version 3 prompts the reader to pause a little longer than indicated by the semicolon in Version 4. Also, Version 3's semicolon suggests a closer relationship between the two clauses than the period does in Version 4.

Yes, the distinctions are rather subtle, but they do exist. Moreover, one of the four versions almost certainly expresses exactly what the reader intended. The comma splice doesn't even come close!

Fused sentences don't even bother with a comma:

> All our networked computers are checked for viruses every month an IT technician carefully searches each computer's hard disk drives.

So what exactly did the writer mean? Did he or she mean to emphasize that all the computers are checked every month, or did the writer mean to say that every month an IT technician searches the hard disk drives? In practical terms, it amounts to the same thing. Or does it? Probably, the IT department would like the emphasis placed on what the IT technicians do, especially if this sentence appears in a proposal to hire more technicians. So there's more at stake here than issues of correctness and rules about when to use periods or semicolons. The real problem with fused sentences is that it's hard to tell what the writer meant.

Misplaced modifiers

Often, a modifying word or phrase can dramatically change the meaning of a sentence by its placement:

> The Act *only allows* engineers to approve designs. [but it doesn't require engineers to do so]
>
> *Only the Act* allows engineers to approve designs. [only the "Act," not any other document]
>
> The Act allows *only engineers* to approve designs. [just engineers, not technologists or foremen]
>
> The Act allows engineers to *only approve* designs. [they can approve but not create the designs]

One of the above versions reflects what the writer meant.

Sometimes, the writer clearly has misplaced the modifier and, although we can decipher what was meant in such cases, the process takes time and the writer loses credibility:

> *Rising from every one of the seven dead trees*, we saw the eagles taking flight. [The eagles rose; we didn't.]
>
> Stéphane Dion delivered a ringing endorsement of federalism, *determined to seem strong in Québec*. [Here, the modifying phrase should immediately precede the subject because Dion is the one who wants to appear strong in *La Belle Province*.]

Dangling modifiers

In some cases, a modifying word or phrase has nothing in the sentence to modify, perhaps with comic effect:

> Jogging along the road, a car swerved and nearly hit me. [Can cars *jog*?]
>
> **Better:** *As I was* jogging along the road, a car swerved and nearly hit me.

Jogging along the road is known as a **dangling modifier**. In the following sentence, a modifier vaguely implies a connection with something or someone not identified in the sentence:

> Speaking with Joe Burns, he said that only the 2007 group of contractors would be considered for 2008 haul licences. [Joe wasn't talking to himself.]
>
> **Better:** Joe Burns told me that only the 2007 group of contractors would be considered for 2008 haul licences.
>
> **Or:** When I spoke with Joe Burns, he told me that....

Disruptive modifiers

Inserting a modifying phrase between a subject and verb, or between parts of a verb phrase, interrupts the natural progression from subject to verb that readers expect:

> Campbell expected the protestors to *fairly quickly* go home.
>
> Workers will, *if they are abused too long*, rebel in one way or another.
>
> The transits, *because we've moved to GPS readings*, were discarded last fall.
>
> He bought *with his first paycheque* a secondhand car.

Instead, place the phrase either before the subject or after the verb:

> Campbell expected the protestors to go home *fairly quickly*.
>
> *If they are abused too long*, workers will rebel in one way or another.

Faulty coordination

Using conjunctions such as *and*, *but*, *so*, *or*, *for*, and *nor* to join two independent clauses works well as long as the two clauses are closely related or equivalent as in the following examples:

> As an engineer he is responsible for ethical designs *and* he takes that responsibility seriously.
>
> The timing belt broke *and* the engine immediately stopped running.
>
> Most engines will simply stop running when the timing belt breaks, *but* "interference" engines will experience major damage when a moving piston strikes an open valve.

The first example contains two clauses closely linked by their emphasis on an engineer's responsibility. Example 3 also describes a close relationship between the two clauses. The second example shows a cause-effect relationship between the two clauses.

These examples of faulty coordination neither express a cause-effect relationship nor a close thematic connection. In each case, the writer needs to choose which clause is subordinate.

Faulty:

Each year the creek overflows its banks and the creek has several sizeable sand bars.

This control box is connected to each sensor and it sends information about each room to the building's main climate control computer.

PharmaHeal Inc. is a private company in Cambridge, Ontario and it started in 2001.

Better:

Each year the creek, which has several sizeable sand bars, overflows its banks.

This control box, which is connected to each sensor sends information about each room to the building's main climate control computer.

PharmaHeal Inc., a private company in Cambridge, Ontario, started in 2001.

Or:

PharmaHeal Inc., which started in 2001, is a private company in Cambridge, Ontario.

A private company in Cambridge, Ontario, PharmaHeal Inc. started in 2001.

Faulty parallelism

Balanced sentences join matching grammatical forms:

The timing belt *broke* and the engine immediately *stopped* running. [*Broke* and *stopped* are **parallel words**, both of which are past tense verbs.]

Short sentences *reduce word counts* and *improve readability*. [The two **verbal phrases use parallel** phrasing.]

As the head engineer, *he oversees all the designs* and *he takes that responsibility* seriously. [The two **clauses use parallel phrasing**.]

Readers may be confused when equivalent words, phrases, or clauses are not phrased in parallel ways:

The proposed upgrades include

- *upgrading* the electrical service from 15 amps to 30 amps;
- *replacement and an upgrade* of the water pipes on the property; and
- *an upgrade of* the existing drain field is anticipated.

The list helps readability, but the italicized words in the three listed items should use the same grammatical form. In this case, the present participle would work well:

The proposed improvements include

- *upgrading* the electrical service from 15 amps to 30 amps,
- *replacing and upgrading* the water pipes on the property, and
- *upgrading* the existing drain field.

Here's another list that requires parallel structure:

> Selecting the priority watersheds to be studied focused on two main considerations:
>
> - *significance of the watersheds* (*high fisheries* or *domestic water* supply values), and
> - *restoration projects must have* a high probability of success.

Yes, the reader can eventually figure out what this relatively short list entails, but that task would be easier if both the bulleted items (and the bracketed phrase) were parallel and if the two bracketed adjectival phrases were also phrased in parallel:

> Selecting the priority watersheds to be studied focused on two main considerations:
>
> - *the watersheds must be significant* (they must have *high fishery* or *domestic water* supply values), and
> - *proposed restoration projects must have* a high probability of success.

Parallelism is also important within standard sentences. The following lack parallel structures:

> Getting to the interview stage can sometimes be more satisfying than to do well at the *interview* itself. [The parallel phrases should be *getting to* and *doing well.*]
>
> Differing data gathering methods *can not only result in* spotty data collection but *also lead to* difficulties in interpreting the data. [The parallel phrases should be *not only result in* and *can also lead to.*]

Practice session 1—identify subjects and predicates

Underline the complete subject in each of the following sentences and circle the simple subject.

Double-underline each complete predicate and circle each simple predicate. You will find answers for these exercises on page 124.

1. You will need to return that spectrometer to Gordon Burns.
2. Will the ratepayers pay the development costs for the proposed riverside trail system?
3. Some air toxins come from natural sources such as forest fires and soil erosion.
4. Air toxics pose a serious threat to human health in all parts of the world.
5. The ozone layer in southern Canada thinned by an annual average of about 6 percent in the late 1990s.

Practice session 2—identify sentence types

In the space provided, name the type of sentence for that item (simple, compound, complex, compound-complex). You will find answers for these exercises on page 124.

6. Although the ozone layer is an important defence against UV penetration, several other factors can have an effect.
7. The air is cleaner and cooler in the Arctic region.

8. Both greenhouse warming and the thinning of the atmospheric layer result partially from human activities, but those activities have had less impact than natural phenomena, whose impact has not been recognized by the orthodox scientific community.

9. Government agencies have focused on gathering data and lobby groups have used that data to push their own agendas, while the oil industry carries on like there's no tomorrow.

10. The Kyoto Accord will not survive in its present form.

Grammar at Work: write short, crisp sentences

"Avoid the pitfall of using excessive technical jargon or complex sentences. Keep the sentences short, crisp and simple. Also, don't jam too much thought content into one post. Instead, add another post later."

—*"Common Mistakes to Avoid," posted June 19, 2007 at the Blogging Strategies website, http://www.bloggingstrategies.com*

Practice session 3—identify sentence errors

Underline each example of a comma splice (CS), fused sentence (FS), dangling modifier (DM), misplaced modifier (MM), sentence fragment (frag), disruptive modifier (DisM), faulty coordination (coord), or faulty parallelism (//) in the following sentences. In the space above the underlined error, write the appropriate symbol (CS, DM, etc.) You will find answers for these exercises on pages 124 and 125.

11. The Ekati mine has cost Dia Met about $536 million and is 70% complete.

12. When grading the site, two-way radios, which allow surveyors to communicate from a distance, are used.

13. Most people like to relax after a hard day's work in front of the TV.

14. After completing the project under budget, the head engineer rewarded me with a promotion.

15. The employee persuaded his boss to grant a pay raise with logical arguments.

16. Many municipalities created, for the benefit of local farmers, policy that provided for minimal monitoring of animal waste runoff.

17. The Integration comparator is also quite popular, it is used most commonly in digital multimeters.

18. When manufacturing a car chassis, materials need to be strong and lightweight.

19. The objectives of this assessment were to ascertain whether the completed work is in compliance with the prescriptions and that the drainage along the road might have been improved.

20. The lake will be temporarily drained during the drawdown it will then increase in size during the re-wetting phase of the project.

Improvement exercises—sentences

The following 30 sentences may contain errors in sentence structure, pronoun usage, adjectives and adverbs, conjunctions, prepositions, spelling, or punctuation. Rewrite those sentences, correcting the errors. If a sentence is already correct, write ***C*** in the space following the sentence. You will find answers for these exercises on pages 125 and 126.

1. Storm water pond construction did not cause negative environmental impacts and was completed following best management practices.

2. The plan area includes about 390 ha and is bound by Powers Creek to the east, Highway 97 to the north, the District of Peachland to the west, and Okanagan Lake to the south.

3. If the native trees are lost or damaged during the construction process, we'll plant three trees for every one that needs replacing. This will ensure an 80 percent survival of shrubs and a 100 percent survival of trees after three years.

4. Watering of theses plants twice weekly for the first two weeks, reduced to once a week for four weeks thereafter, weather dependant, will ensure quick root growth.

5. Myself and all the team at ESB look forward to hearing from you.

6. One of his finer projects was the Penticton Channel recreation area, he helped design and he oversee the construction of this project.

7. This proposal outlines the proposed research that will need to be conducted, as well as, the solution to the problem that we will implement.

8. By erecting a breakwater wall, the currents will be diverted, thereby eliminating beach erosion.

9. Facial expressions are raising or lowering the eyebrows, squinting, clenching the jaw, as well as many others.

10. pH measures the acidity or alkalinity of water and many organisms that live in the water are sensitive to changes in pH.

11. These steps, keep sun exposure to a minimum, especially between 10 a.m. and 3 p.m. and wear wide-brimmed hats and wear sun screen with Sun Protection Factor of 15 or greater and reapply the sunscreen frequently, will help protect you from the sun's harmful rays.

12. Fearing the dilution of its trademark, remembering what happened to brand names like Kleenex and aspirin and trampoline, and seeing the word entrenched in dictionaries such as the Oxford English Dictionary, and noting the increasing popularity of the word, Google Inc. tried to persuade members of the public to refrain from using "google" as a verb.

13. A post-development/construction report summarizing the development, outlining the degree of compliance with the above measures and demonstrating that the SPEA is protected will also be produced and submitted on the RAR database.

14. The proposed "F"-shaped dock will be made of perforated material and consists of a 1.8 m wide mainstem.

15. None of the species are recently reported in the area, and it is not likely that the storage of wood waste on the subject property will significantly affect any of these species.

16. The average person spends 3,220 hours stuck in traffic, but only minutes planning their retirement.

17. Client A has not attended regularly, nevertheless, he continues to make contact with our counselors.

18. Smoking is only allowed in the gazebos at this campus.

19. Because each of us has a responsibility to the organization.

20. Trying their very best in the aptitude tests, we observed the clients at the Centre.

21. If you know the answer, it is easy, however, not everyone knows the answer.

22. After sending the message, the following screen will appear.

23. Specialty NiCd batteries have a niche market in the area of cordless and wireless telephones, emergency lighting, model airplanes, as well as power tools.

24. With a relatively low internal resistance, a NiCd battery can supply high surge currents. This makes them a favourable choice for remote controlled electric model aeroplanes, boats, and cars, as well as cordless power tools and camera flash units.

25. This workshop series has three objectives, including

    - the development of a strategies for structuring documents.
    - learning how to proofread and edit other's documents
    - as well as improving ones' efficiency in writing.

26. Please feel free to contact myself if you require further clarification.

27. This is the first tier of the assessment process and is completed by the supplier.

28. Roughly 100 pharmaceuticals have now being identified in rivers, lakes, and coastal waters throughout Europe and in north America. In concentrations that have alarmed environmental protection agencies.

29. The European medicines Agency recommend additional testing of drugs to learn how they break down when they enter surface or underground water supplies.

30. One worry for pharmaceutical companies is that increased amount of testing could costly translate into delays for the release of new drugs.

Document critique—sentences

The following email may contain errors in sentence structure, grammar, capitalization, punctuation, or spelling. Underline all errors and identify them using standard proofreading marks. (See the Correction Symbols list on page 144144.) You will find answers for this exercise on page 127.

Note: Do not comment on the tone of the email; simply identify errors.

Corrine Maltais

From:	Billie Cochrane <bcochrane@on.gc.ca
To:	Corrine Maltais <cmaltais@on.gc.ca
Sent:	Friday, June 27, 2008 7.48 AM
Subject:	Conflicting responsibilities for The Georgian bay Project

As my project leader, I hope your can answer my question about what Im suposed to be doing on the boat. When I took this on as my summer student placement, I thot that myself and the other student wood just be observing procedures and collecting data, not doing all the grunt work on the boat. Isnt that what sailors are for?

Here's what been happening –

- On Monday, myself and Jeannie (shes the other student) had to coil all the ropes and check for flaws in the ropes, that's very hard work, let me tell you, those ropes are heavy and some of them are dirty too.
- Then on Tuesday the captain made us clean the decks, he said that were the ones getting the deck dirty with all our fish samples and because we came on board with muddy boots after checking the mud flats at 3 Moon bay. That's a big boat actually and we had to use mops and then hand scrub with those hard sponge things. Why doesn't they have proper motorized cleaning equipment on board?
- Wedensday we sailed to parker Point to get water samples and to check for fish mortality. Remember we had heard reports that fish were washing up on shore we didn't see any though, maybe the reports were for another point. That was a good day, we collected samples from 17 lakeshore locations and 5 deep water locations we observed three outflow pipes and took samples from those outflows, one of them I think it was the pulp mill around the corner from Parker Point was really disgusting – the colour was muddy red and it smells like sulphur and diesel fuel which is a really foul combination for sure.
- Yesterday as you know we spent in the lab with your techs as they checked the samples for toxic substances they found very high levels of dieldrin and lindane which was surprising because I thot that those pollutants were in decline, at least that's what Professor Gorsline said.

- This morning Jeannie and myself went down to the boat to retreive some equipment and the captain said that we had to clean the cabin and the storage lockers down below. Whoa I said, that's not what I came here for. Im a forth year biology student, I said, and Im gonna tell Dr Maltais whats going on and he said go right ahead I don't care but its my boat and youll do what I tell you when your on board That's what he said. Well I came right up to see you but you were in a meeting so Jeannie said I better write to you right away.

PUNCTUATION

On a recent drive through a Montana mountain pass, I grudgingly followed a truck whose back doors screamed, "Car's RV's Boats We repair Them All!" in letters two feet high. What possessed the perpetrator of this crime to place the apostrophe in the plural form of *car*? And why, I wondered through clenched teeth as I followed the truck through swirling snow, were *them* and *all* capitalized? This type of response to faulty punctuation appears in the witty, ruthless expose of mismanaged punctuation in Lynne Truss's book, *Eats, Shoots & Leaves*.[5] Truss makes a case for paying close attention to the punctuation we employ:

> ... all our thoughts can be rendered with absolute clarity if we bother to put the right dots and squiggles between words in the right places. Proper punctuation is both the sign and the cause of clear thinking. (202)

Punctuation marks were introduced to the written language during the age of printing by typesetters who wanted to approximate the pauses, tonal emphasis, and musical pitch that we easily use when we speak. In the intervening half-millennium, punctuation methods and styles have changed, but the key punctuation marks have survived. This chapter examines periods, commas, colons and semicolons, question marks and exclamation points, apostrophes, dashes, hyphens, italics, and underlining.

End Punctuation

Conventional usage places a period, a question mark, or an exclamation point at the end of a sentence:

> That laptop is yours to use this semester.
>
> Is that your laptop?
>
> Your laptop is amazingly light!

Colons and semicolons often appear at the end of independent clauses as in the following examples:

> I've ordered a laptop that has all the features you requested:
>
> - a 17" widescreen LCD,
> - an MP camera,
> - 3 GB RAM,
> - a 300 GB hard disk drive, and
> - a dual processor.
>
> The laptops will be here Thursday; please install networking software as soon as they arrive.
>
> You'll need a wireless router: an SMC Barricade 2804BR would fit your system.

---

[5] Lynne Truss, *Eats, Shoots & Leaves* (London: Gotham Books, 2003).

Colons and Semicolons

Colons point at things, as if to say "Here, look at these words." As the examples at the bottom of page 62 illustrate, colons can appear at the end of independent clauses. However, a colon should not be placed at the end of an incomplete thought, such as a phrase or dependent clause.

> If we have a labour stoppage next week, do not enflame the situation by actions such as
>
> - crossing the picket lines,
> - sharing your feelings about the strike with reporters, or
> - conducting classes at a location outside the College.

Hints:

When the lead-in words at the end of an independent clause are *including*, *consists of*, *like*, or *such as*, do not use a colon. You can use a colon if the lead-in words are *as follows* or *the following*.

You can use a colon to introduce a **quotation** or an **example**:

> Truss believes we should resist the erosion of standard punctuation: "The reason to stand up for punctuation is that without it there is no reliable way communicating meaning" (20).
>
> Some punctuation marks are overworked: for example, Lynne Truss lists seven discrete uses for the apostrophe.

After an independent clause, you can use a colon to introduce an **appositive**, a word or collection of words that renames a noun or pronoun.

> The Mitsubishi i-car has a major competitive advantage; the design of this diminutive vehicle allows four passengers to ride in comfort.

Semicolons also have several uses:

- A semicolon can separate independent clauses if the writer wants to show that the two ideas are closely related (as shown by a pause rather than by a full stop):

  > Whirlpool Corporation has been busy acquiring brand names; its stable of brands now includes Amana, Whirlpool, Maytag, Inglis, Admiral, Jenn-Air, Magic Chef, and KitchenAid.

- If the second independent clause starts with a conjunctive adverb or transitional expression, a semicolon could be placed between the two clauses:

  > The Maytag dishwasher has a distinctive, independent brand name; however, its replacement parts come directly from Whirlpool.

- When each of the items in a series is comparatively long, or if each item contains commas, the items might be separated by semicolons rather than commas:

  > The Canadian Olympic Committee's responsibilities include athletic, cultural, and educational programs promoting Olympic values in Canada; grassroots programs in communities across the country; and selection of Canadian cities to host Olympic, Olympic Winter, and Pan American Games.

Commas

In speech, pauses convey a wide range of nonverbal meanings. Commas attempt to convey the same meanings in written messages; however, it's often difficult to decide where to place commas.

As the above case illustrates, sometimes that placement has dramatic consequences. Hopefully, the following list of uses will help you decide when and where to employ commas.

Commas follow introductory elements:

- An introductory **word**: *Hopefully*, the rain will hold off until tomorrow.
- An introductory **phrase**: *After the winter rains*, the desert flowers burst out in full glory.
- A **transitional expression**: *For instance*, three soldiers have stepped on landmines this week.
- A **clause**: *Even though the bridge passed a recent inspection*, a jittery provincial government has closed all bridges of this design, pending further examination.

Commas separate adjectives that contribute independent of each other:

Every *tiny, feature-packed, 4 GB* MP3 player in the display case was on sale.

Commas <u>do not</u> separate adjectives that contribute together:

The *dirty white* van is still parked in front of our house. [The van is not white; it's *dirty white*.]

Really, he's a *strange old* man. [There's *old*, and then there's *strange old*!]

Commas precede nonrestrictive elements (elements that are not essential to the meaning):

The Rogers/Aliant comma controversy*, which cost Rogers millions,* was totally avoidable.

Commas <u>do not</u> precede restrictive elements (elements essential to the meaning):

I was very interested in the Rogers/Aliant comma controversy *that the Globe and Mail reported.*

Commas separate items in a series (three or more elements constitute a *series*):

Recent Antarctic ice breakups may affect sea levels, ocean temperatures, and Antarctic weather patterns.

Creating bronze statues requires specialized equipment, skill, patience, and expensive materials.

The practice of burning yard debris in suburban neighbourhoods has implications for people with breathing problems, for bylaw officers, and for those responsible for municipal landfills.

Note that a comma appears before the coordinating conjunction (*and*, *or*) that precedes the last item in a series:

Health problems might include eye irritation, headaches, *asthma, or bronchitis.*

If a comma does not appear before the coordinating conjunction in the above sentence, *asthma or bronchitis* can be interpreted as a single phrase, not as two separate items.

A comma <u>does not</u> appear within an item that incorporates *and*:

You have a choice of an omelette, *ham and eggs*, or a Denver sandwich.

Commas may set off parenthetical expressions:

Residents of Sooke (a suburb of Victoria, B.C.) have complained about backyard burning.

Residents of Sooke*, a suburb of Victoria, B.C.,* have complained about backyard burning.

A total of $148 million in capital spending, *mostly housing projects,* is planned for the spring.

Commas may set off quoted words from explanatory or lead-in words:

Ross Grieve, CEO of PCL Constructors Inc., *said,* "*Employee* stock ownership is a good way to compensate people and tie them to the company."

However, a comma is not needed when the quoted words blend into the grammatical structure of a sentence:

> Ross Grieve believes that *"Employee* stock ownership is a good way to compensate people and tie them to the company."

Commas may appear in dates:

> March 28, 2008 Friday, March 28, 2008

Commas don't appear when they don't have to separate numbers:

> March 2008 28 March 2008

Commas with numbers:

The metric system doesn't require commas in numbers that have more than four digits:

> 98 429 124 654 190

However, many people still use commas:

> 5,397 98,420 124,654,190

Commas with names, places, and addresses:

Here are some examples of comma usage with names, places, and addresses:

> Maxwell Sandestin, M.D.
>
> Saskatoon, Saskatchewan
>
> I sent the parcel to 26 Prince Andrew Place, Don Mills, Ontario M3C 2T8. [Notice the comma usage when an address is part of a sentence.]

Do not use a comma in a name such as *Martin Luther King Jr.* or *Davis Love III.*

Commas in correspondence:

Emails and informal letters place a comma after the salutation:

> Hello Mr. Konecsni,
>
> Dear Judy,

Formal letters use a colon after the salutation, or after an attention line:

> Dear Ms. Ramage:
>
> Attention: Ms. Frances Ramage

Complimentary closes in letters or formal emails require a comma:

> Yours truly,
>
> Sincerely,
>
> Best regards,

Hints for Inserting Commas:

Inserting a comma where <u>you</u> would pause during speech provides one way of knowing when to use a comma, but not everyone thinks or breathes in the same patterns, so equating commas to your speech pauses is not necessarily useful. A better strategy is to place a comma where you would like your <u>reader</u> to momentarily pause.

If you're unsure, circle the point where you think a comma might be useful, and later check the advice provided in a grammar handbook. (And remember that *Grammar and Style at Work* provides only that advice needed to guide you in this book's exercises—the 2½ pages of advice and examples in this workbook do not offer the level of detail that you'll find in a grammar handbook.)

Apostrophes

The apostrophe came into English in the 16th century, when its sole purpose was to stand in for dropped letters (as in Shakespeare's use of *'tis* instead of *it is*). Since then, it has been given much more work to do, starting in the 17th century when printers inserted the apostrophe in singular possessive nouns (such as *the printer's mistake*).

These days, the apostrophe indicates the following kinds of possession:

- **Possessive form of singular nouns**

  John's wedding the duck's wing Sandra's banquet Ross's car

- **Possessive form of plural nouns**

  printers' conventions politicians' indiscretions the Browns' house

  the Joneses' house the Rosses' car

- **Possessive form of compound words and phrases**

  my father-in-law's response Member of Parliament's expense account

- **Possessive form of individual nouns in a phrase**

  John's and Karla's wedding ringsthe student's and the professor's interpretations

- **Form of the last noun in a joint or group possession**

  John and Karla's house John and Karla's two dogs

- **Form of possessive indefinite pronouns**

  No one's phone has been tapped. Someone's phone is ringing.

Note:

Do not use apostrophes with possessive pronouns such as *my*, *your*, *hers*, *his*, and *its*.

The apostrophe also indicates when one or more letters have been omitted in **contractions**:

aren't can't didn't don't I'm she's

Apostrophes are <u>not</u> used for the possessive forms of pronouns such as *my*, *mine*, *your*, *yours*, *his* *hers*, *its*, *our*, *ours*, *their*, *theirs*, and *whose*. So do not confuse the following possessive pronouns and contractions that sound alike:

Possessive pronoun	Contraction
its	it's (it is)
your	you're (you are)
their	they're (they are)

whose who's (who is)

Recent practice has discouraged using the apostrophe in forming **plurals of numerals, years, acronyms, and symbols**. The following sentences illustrate current usage as defined by the Modern Languages Association:

How many *8s* appear in that sequence of numbers?

In the early *1900s*, very few motor cars were evident in Canada.

At last count, 237 *MPs* favoured a major overhaul of the income tax system.

I use very few *@s* and *&s* in my writing.

Dashes

Dashes let you interrupt sentences to add information or to temporarily take the sentence in a different direction. A dash can operate by itself, but more often dashes work in pairs. A dash is two or three times the length of a hyphen; word processing software such as MS Word will automatically lengthen a hyphen into a dash when you add a space and a word after the hyphen.

Here are some examples of dashes in action:

Canadian NHL teams—Montreal, Toronto, Edmonton, Calgary, Vancouver—have all welcomed the strong Canadian dollar.

What could be more appropriate*—or more profitable—*than boosting the provincial royalties on oil and gas production?

Her P.R. best work*—and I say this in all sincerity—*has resulted from happy accidents, not from careful planning.

Hyphens

Hyphens have several applications:

- They separate compound words such as the following:

  self-assured president-elect well-dressed three-year plan anti-government

  However, they <u>do not</u> separate compound words in phrases such as the following:

  poorly managed project most appropriate solution a half week's work

  a more sensible approach

- They appear after the prefixes *all-*, *ex-*, *quasi-*, and *self-*:

  all-inclusive ex-army officer quasi-official self-employed

  They <u>do not</u> appear after the prefixes *re*, *un*, *and inter*:

  revamp unconventional interplay

- They are used when a prefix or suffix is added to a number or to a proper noun:

  post-2001 pre-Raphaelite Harper-like pro-Canadian

- They are used when the vowel at the end of a prefix or suffix is the same as the first letter of the root word:

  re-entry anti-inflammatory

Parentheses and Brackets

Like dashes, parentheses allow you interrupt a sentence to insert information. However, dashes emphasize surrounding statements, while parentheses provide interesting asides, examples, or short explanations. Here are some examples:

> This grammar workbook *(the first of its type)* is designed to supplement the 6th Canadian edition of *Technical Communication*. [an aside]
>
> Other grammar workbooks (*such as Grammar Unplugged*) focus on exercises, with little advice or direction about how to complete the exercises. [an example]
>
> This workbook has been printed in black and white (to save costs) in an 8½ × 11 inch format (to accommodate the exercises). [explanations]

Legal writing often places parentheses around a numeral that follows a spelled-out number (*for a period of five (5) years*), but technical writing does not use this practice.

Brackets (also known as "square brackets") have very few uses, mainly in academic writing. Academics use brackets to insert words into quotations. For example, here's a quotation used in this workbook:

> "Everyone makes mistakes, but a 'minor' slip-up in writing—such as the start date of an important conference—can cost companies time, money and credibility."

Here's how one might incorporate part of that quote into a sentence:

> Author Ronnie Moore notes that "Everyone makes mistakes, [that] can cost companies time, money and credibility." [The conjunction *that* is inserted to help the sentence flow smoothly.]

You've no doubt noticed that this book uses brackets to insert explanatory notes following certain examples, as in the one above.

Brackets are also used to enclose the Latin term *sic* when you want to inform the reader that an error in a quote is in the original document you're quoting:

> The review board was uncertain about its' [*sic*] role in resolving the dispute. [Like all foreign words, *sic* is italicized in the above example.]

Underlining and Italics

In the good old days of typewriters, **underlining** was used to highlight the names of books and other publications. Headings were also underlined. In the age of electronic word processing, those functions have been superseded by italics. The only job left for underlining these days is to emphasize words or phrases:

> The earth's ozone layer is in <u>very deep</u> trouble.
>
> The earth's <u>stratospheric</u> layer of ozone protects us from the sun's harmful ultraviolet rays.
>
> The "ozone hole" refers to <u>a large and rapid decrease</u> in the numbers of ozone molecules.

As you can see in this book or in any book, magazine or newspaper, **bold type** has replaced underlining for headings. Meanwhile, **italics** have taken over many of the other tasks once assigned to underlining. Here are some of the uses for italics.

- Italics grace the names of ships, airplanes, and trains:

the *Orient Express* *Titanic* *Spirit of St. Louis* [the plane flown by Charles Lindbergh in the first transatlantic flight]

- Italics signal the titles of books, magazines, newspapers, films, television programs, CDs, and websites:

  Technical Communication *Maclean's* *Globe and Mail* *Corner Gas*

  Music for the Hard of Thinking *The Weather Network*

- Italics accompany foreign words and the scientific names of plants and animals:

  laissez-faire *ad infinitum* *canis lupus* *iris versicolor Linné*

- Italics are used for words, letters, and numbers referred to as words:

  Words like *groovy* and *cool* no longer have the in-crowd clout they once had.

  Here's a hint: the letter *s* appears in the word four times.

  Out of respect for the great players who had made the numbers *9*, *99*, and *66* famous, Crosby chose *87*.

- Italics can also be used to emphasize certain words:

  None of his designs could be considered great work, *except* for his work on the Ford Mustang.

Caution:

If you use italics for purposes other than emphasis in a document, you should not use italics to emphasize certain words. Instead, you should underline the words or use boldface.

Quotation Marks

Quotation marks enclose short bursts of quoted material, which is one of several uses for these marks. The following descriptions include advice and examples:

- Quotation marks surround words that are quoted exactly as they appear in the original document:

  Lynne Truss believes that the comma has become "a kind of scary grammatical sheepdog."

  She also says that, "As we will shortly see, the comma has so many jobs … that it tears about on the hillside of language…."

  Note that the first sentence leads into the quotation without a comma and starts the quotation with a lowercase letter, because Truss's partial sentence fits into the flow of the containing sentence. The second quotation is preceded by a comma and begins with a capital *A*, which is standard practice for quoting a full sentence. Note also that part of that second quotation has been deleted, and the omitted words have been replaced by **ellipses points** (…).

- Quotation marks do not enclose a longer quotation, which is sometimes defined as 50 words or longer (APA style) and sometimes identified as longer than four lines (MLA style). These quotations are indented and single-spaced, and do not use quotation marks:

Lynne Truss makes a fierce case for paying attention to punctuation:

> But after journeying through the world of punctuation, and after seeing what it can do, I am all the more convinced we should fight like tigers to preserve our punctuation, and we should start now.

> Who wants a blank map, for heaven's sake? There is more at stake than the way people read and write. Note the way the *Washington Post* news story explained the benefits of emailing: it "increased employees' productivity by 1.8 hours a day *because they took less time to formulate their thoughts.*"

The above quotation can use quotation marks within the quoted passage to denote words that Lynne Truss is quoting. However, if your short quote cites an internal quotation, you need to use the following technique, which employs regular quotation marks for the full McCourt quote and partial quotation marks for the bit he quotes:

> McCourt cites Lynne Truss's combative spirit: "Then she flings down the gauntlet: '... he would be ill-advised to repeat this ploy once my punctuation vigilantes are on the loose.' (Notice my masterly use of the ellipsis....)." Clearly, McCourt admires Truss's feisty nature.

Right! Had enough? If so, proceed to the following exercises.

Practice session—punctuation

Circle the letter of the preferred choice in the following pairs or groups of sentences. You will find answers for these exercises on page 128.

1. **a.** Three persons attended the conference: Garth Barton, shipping and receiving department, Marcie Jones, Marketing Division, and Rita Kohlmann, general manager.

   b. Three persons attended the conference: Garth Barton, shipping and receiving department, Marcie Jones, Marketing Division, and Rita Kohlmann, general manager.

2. **a.** Our biggest challenge is: how to increase the bore size while retaining wall strength.

   b. Our biggest challenge is how to increase the bore size while retaining wall strength.

3. **a.** Attention, Mr. Francis

   b. Attention: Mr. Francis

   c. Attention—Mr. Francis

4. **a.** Four weeks is a long time to wait for the test results: Our competition might win the mineral rights in the interim.

   b. Four weeks is a long time to wait for the test results, our competition might win the mineral rights in the interim.

   c. Four weeks is a long time to wait for the test results; our competition might win the mineral rights in the interim.

5. **a.** I wonder, should we hire the experienced candidate?

   b. I wonder: should we hire the experienced candidate?

6. **a.** Mel's and Doug's laptops both need new hard drives.

   b. Mel and Doug's laptops both need new hard drives.

7. **a.** We are able to supply desktop computers, servers, monitors, printers, and maintenance and service.

   b. We are able to supply desktop computers, servers, monitors, printers, and maintenance, and service.

8. **a.** Do you still watch Hockey Night In Canada, or do you find it boring?

   b. Do you still watch <u>Hockey Night In Canada</u> or do you find it boring?

   c. Do you still watch *Hockey Night In Canada*, or do you find it boring?

9. **a.** No one notices when we do excellent work, I wonder if they know we're here.

   b. No one notices when we do excellent work I wonder if they know we're here.

   c. No one notices when we do excellent work; I wonder if they know we're here.

10. **a.** Grammar used to be my most hated subject but that was before I took this course.

    b. Grammar used to be my most hated subject, but that was before I took this course.

    c. Either **a** or **b** would be acceptable; your choice is a mater of preference.

Improvement exercises—punctuation

The following sentences may contain errors in grammar, punctuation, capitalization, and spelling. Underline each error and, in the space provided, correct the error. If a sentence contains no errors, write ***C*** in the space provided. You will find answers for these exercises on pages 128–130.

1. The Trackstick has over 1 Mb of memory it can store months of travel information.

2. The Trackstick can record your exact routes, when you bike, hike or run.

3. The Trackstick has several applications, employee monitoring, vehicle tracking, speed control, photo tours, public safety, law enforcement

4. CEO Hernan Hesse says that the Trackstick II offers: "Upgraded features and accuracy compared to the original device.

5. The device receives signals from 24 orbiting Satellites, that precisely calculates its position on this planet.

6. The property is owned by Northern Tent Vacations, and the president carlo DeValero, the Company has authorized Green Consultants to make applications, for environmental approvals as required for development

7. Don't worry about the condition of it's transmission; the rest of the car is in really good shape.

8. The proposed wharf will be the same length and appearance it will use the same materials as the the adjacent wharf.

9. She orchestrated, and wrote, a major update of a maintenance manual.

10. Shortly after he graduated with a engineering diploma he develops a process for evaluating the quality of concrete as it is poured.

11. Prominent scientists, such as Henri Casselle have questioned the popular beliefs about global warming.

12. Art Burns' designs won the Granby Award for high density housing projects this year.

13. Well, their the ones who own the fishing rights so we should back off.

14. The Canadian dollar—as strong as it is has hurt U.S. exports that are paid in U.S. dollars.

15. The Leafs post 2005 regular season records leave a lot to be desired.

16. The rocket burst into flames upon reentry into the earth's atmosphere.

17. Generation Y workers are not all that different from any other "generation" of workers; according to workplace analyst Barbara Moses.

18. This grammar workbook [which supplements your *Technical Communication* textbook] provides hundred's of exercises.

19. The Spanish say, hasta mañana and thats' exactly how I feel about wanting tomorrow to be here now!

20. On 10 June, 2004 a British Airways Boeing 777 suffered a serious fuel leak as it took of from heathrow airport.

21. Lack of documentation [or poor documentation] costs your company money.

22. The total cost of a single call answered by a help desk technician is, $20 or more depending on the length of the call.

23. Innovatia reports on the costs of distributing poorly written documentation on their website.

24. The Regional Aquatics Monitoring Program (r.a.m.p.) is an industry funded multi-stakeholder initiative.

25. The Athabasca river has high year to year fluctuations in its flow patterns.

26. “I’d be shocked if we learned that the cause of the flooding was one of the valves installed last week” Said Roger Desroches Managing Director.

27. Unconscious bias may cause some doctors to more likely recommend knee replacement surgery to men than to women according to a recent study.

28. Compared with men, women have the surgery when their arthritis is at a more advanced stage, this perhaps explains why women tend to do more poorly than men after knee replacement.

29. The water quality in the Athabasca river in the fall, 2005, was better than in the previous year.

30. On March 25 2008 the Globe and Mail carried a story about Biovails involvement in a series of frauds that distorted reports of earnings, and revenue.

31. The US Securities an Exchange Commission said Biovail executuives were “obsessed with meeting quarterly and annual earnings guidance” and, engaged in a “corrupt strategy” to do so.

32. In the United States Sirius Satellite Radio Inc. has taken over its rival XM satellite radio holdings inc. which means that the Canadian subsidiaries will also merge.

33. Montreals waterfront district once a bustling scene of shipping and manufacturing has almost become a ghost town however current plans for private and public investment promises to rejuvenate the area.

34. The Canadian fashion industry is not centered in one location rather its found in Winnipeg, Montreal Toronto and Vancouver and of all places in Vernon British Columbia, the home of Far West and Valhalla Pure.

35. Several media networks including traditional media companies like Forbes Inc. are trying to lure advertisers away from google inc. and yahoo Inc. by offering onestop shopping for ad placement.

36. Each of the consultants wrote a report about Riverbend Dams projected cost overruns

37. In this book the term "grammatical" refers to effective well formed phrases and sentences.

38. Rollover rates for SUVs are nearly twice that of sedans, and 30 percent higher than that of pickup truck's.

39. Brick walls can withstand considerably higher temperatures than concrete board walls.

40. Sundins career as a Toronto Maple Leaf appears to be ending which is really two bad.

41. Don't expect your cellular phone's to receive strong signal's in the far northern Arctic.

42. According to Rene Corbeils report the bridge is no longer safe but we knew that.

43. Blogging has real advantages for company's that possess clear goals and a workforce who's ambitions align with the companys'.

44. Guerilla e-mail [www.guerilla-mail.com] supplies disposable e-mail addresses which expires after 15 minutes so you can register a software program or service but not receive spam later on.

45. A study by York university professor Dr Robert Kozinets shows that consumers of technology are governed partly by their view of the role that technology plays in they're lives and in society.

Document critique—punctuation

The following document may contain errors in grammar, capitalization, punctuation, or spelling. Underline all errors and identify them using standard proofreading marks. (See the Correction Symbols list on page 144.) You will find answers for this exercise on page 131.

Note: The following email responds to an email that appears on page 60 of this workbook.

Billie Cochrane

From: Corrine Maltais <cmaltais@on.gc.ca

To: Billie Cochrane <bcochrane@on.gc.ca

Sent: Monday, June 30, 2008 11.19 AM

Subject: Conflicting responsibilities for The Georgian bay Project

Thank you for drawing my attention to your experiences on the great lake queen our research vessel I have spoke with captain Blair whose denying the accusation made by yourself and Jeannie Brock.

Let me respond to each of youre comments;

- On Monday, you say that captain Blair forced yourself and Jeannie to coil all the ropes and check for flaws. He says that he asked you to help himself and the two sailors but only for 15 minutes that's because they had a particularly difficult time with the anchor rope that somehow become kinked. He said that after you helped with the anchor rope which he appreciated by the way he saw you coiling some dock lines however he says he didnt order you to do that.
- He said that on Tuesday all six of you including himself cleaned the deck.
- On Friday, when you went to retrieve the collection equipment according to captain Blair he asked you to clean the cabin because you had left a number of items lying around during Wednesdays trip to parker point also he claims that you and Jeannie didn't clean up the galley after you prepared lunch, by the way the captain said that him and the crew appreciated you're offer to cook lunch on Wednesday and it was good too he said.

So heres where were at, I think we [you Jeannie myself and the captain] need to get together to discuss your formal roles in the research team, while your on board the great lake queen. We also need to talk about the things that team member do to help each others out and to work as a team – often these things go beyond the formal defined roles you know. Ill set up a meeting for this Wednesday before you sail to The Ramsey shoals. Watch for a email later today which will set the time we should meet on the boat I think.

By the way, you mentioned that your Professor had said that dieldrin and lindane are in decline in the great lakes. Probably his comments were based on a flawed study conducted by the Michigan Fish and Game league. Their results have since been shown to been based on a too small sample and solely on levels of those chemicals found in one fish species.

MECHANICS

Mechanical aspects of written documents— capitalization, abbreviations, numbers, spelling, spacing, graphics, punctuation—quietly help readers navigate pages, paragraphs, and sentences. Spacing, graphics, and headings systems are discussed in Chapter 13 of *Technical Communication*. Punctuation is discussed in this workbook's previous section; this section covers spelling, capitalization, abbreviations, and numbers.

Grammar at Work and Play: spelling does matter!

Tobi Gutt, a 21 year-old German left his German home on December 30, 2006 for Australia, dressed in T-shirt and shorts for the Australian summer and looking forward to seeing his girlfriend. Unfortunately, he had typed in "Sidney" instead of "Sydney" on an online booking form. He ended up in Sidney, Montana, 13,000 kilometres off course, with stops in Portland, Oregon and Billings, Montana.

After three cold days in his summer clothes, he received €600 from home and he was able to book a new ticket to Australia.

Spelling

English includes words from many sources. Here's the result of one attempt to determine the percentages of English words derived from the contributing sources of our language:

> ... the result of a computerized survey of roughly 80,000 words in the old *Shorter Oxford Dictionary* (3rd edition) was published in *Ordered Profusion* by Thomas Finkenstaedt and Dieter Wolff (1973). They reckoned the proportions as follows:
>
> - Latin, including modern scientific and technical Latin: 28.24%
> - French, including Old French and early Anglo-French: 28.3%
> - Old and Middle English, Old Norse, and Dutch: 25%
> - Greek: 5.32%
> - No etymology given: 4.03%
> - Derived from proper names: 3.28%

(Source: ***AskOxford.com***, http://www.askoxford.com/asktheexperts/faq/aboutenglish/proportion?view=uk, accessed April 1, 2008.)

Each one of the above sources brought its spelling conventions and sound combinations, so it's no wonder that English can be difficult to spell. For example, the "eau" in bureau is pronounced the same as "oe" in *hoe* and the same as "ow" in *grow*, and the same as "o" in *holocaust* or *oration.* Those words come respectively, from French, Germanic, Greek, and Latin origins. But words can have the same origin and still require wide variations in the letters chosen to represent sounds. For example consider the "ough" combination, which is pronounced differently in *through* (oo), *trough* (off), *rough* (uff), *bough* (ow), and *bought* (ott)! All five of those words have Germanic origins.

Unlike most other European languages, English does not employ phonetic spelling, despite attempts to introduce "logic" to spelling English. (Noah Webster did have limited success, starting with his dictionary in 1828. Through his efforts, American spelling includes *plow*, *threw*, and *check*, among others.) The real problem with English spelling, though, is that we use a 26-letter alphabet (originally created for Latin) that needs to somehow represent 40 distinct English sounds.

So there's no easy answer for spelling English, especially for Canadians who waver between British spelling (*programme* and *centre*) and American spelling (*program* and *center*). Spell check helps to some extent, particularly if you set the default language to Canadian English. (In Microsoft Word 2010, click on "References/Language/Set Proofing Language ...," which will let you choose from among several variations of English.)

We can teach ourselves the correct spelling of many key words. The following lists are designed to guide your efforts.

Commonly misspelled words

acceptable
accidentally
accommodate
acquire
acquit
a lot
amateur
apparent
argument
atheist
believe
bellwether
calendar
category
cemetery
changeable
collectible
column
committed
conscience
conscientious
conscious
consensus
definite(ly)
discipline
drunkenness
dumbbell
embarrass(ment)
equipment
exhilarate
exceed
existence
experience
fiery
foreign
gauge
grateful
guarantee
harass
height
hierarchy
humorous
ignorance
immediate
independent
indispensable
inoculate
intelligence
its/it's
jewelry
judgment
kernel (colonel)
leisure
liaison
library
license
lightning
maintenance
maneuver
medieval
memento
millennium
miniature
minuscule
mischievous
misspell
neighbor
noticeable
occasionally
occurrence
pastime
perseverance
personnel
playwright
possession

precede
principal/principle
privilege
pronunciation
publicly
questionnaire
receive/receipt
recommend
referred
reference
relevant
restaurant
rhyme
rhythm
schedule
separate
sergeant
supersede
their/they're/there

Commonly confused words

adverse unfavourable, bad	**averse**, strongly disliking or opposed to
assure pledge an outcome	**ensure** make an outcome certain
	insure secure the payment of money
affect make a difference to	**effect** a result (or cause a result)
ambiguous open to interpretation	**ambivalent** means having mixed feelings
amoral not concerned with morality	**immoral** means not conforming to accepted moral standards
appraise assess	**apprise** inform
augur sign of a likely outcome	**auger** a tool used for boring or conveying (as in a grain auger)
censure express strong disapproval	**censor** suppress unacceptable parts of (a book, film, etc.)
climactic forming a climax	**climatic** relating to climate
complacent smug and self-satisfied	**complaisant** willing to please
complement enhance by extra features	**compliment** an expression of praise or congratulation
continuous without interruption	**continual** happening frequently, with intervals between
council an administrative or advisory body	**counsel** advice or guidance
councillor a member of a council	**counsellor** counsels re personal or psychological problems
credible believable, convincing	**creditable** deserving acknowledgement and praise
definite certain, sure	**definitive** decisive and with authority
defuse remove a fuse or reduce the danger or tension	**diffuse** spread over a wide area
desert a waterless area	**dessert** the sweet course
discreet careful not to attract attention	**discrete** separate, distinct
envelop wrap up, cover, or surround completely	**envelope** used to enclose a letter or document
fawn a young deer, light brown colour	**faun** a Roman deity that is part man, part goat
grisly causing horror or revulsion	**grizzly** from *grizzled* ; refers to the bear's white-tipped fur
hoard a store of something valuable	**horde** disparaging term for a large group of people.
its possessive form of it	**it's** short for either *it is* or *it has*

loath reluctant, unwilling	**loathe** dislike greatly
loose unfasten or set free	**lose** means cease to have or become unable to find something
luxuriant rich and profuse in growth	**luxurious** very comfortable and extravagant
marital of marriage	**martial** of war
ordinance an authoritative order	**ordnance** guns or munitions
palate the roof of the mouth	**palette** an artist's board for mixing colours
pedal move by means of pedals	**peddle** a verb meaning sell (goods)
principal first in order of importance; main	**principle**, a basis of a system of thought or belief
proscribe condemn or forbid	**prescribe** issue a medical prescription; recommend an action
stationary not moving or changing	**stationery** paper and other writing materials
story a tale or account	**storey** is a floor of a building
tortuous full of twists and turns	**torturous** means characterized by pain or suffering
who's contraction of *who is* or *who has*	**whose** possessive form of *who*
you're contraction of *you are*	**your** possessive form of *you*

Commonly misused words and phrases

alot There's no such word; use *much* or many. However, there is a word spelled *allot*, which means assign, allocate, distribute. *A lot* is considered slang.

could of The phrase should be *could have*. (Use the helping verb *have* with *would*, *should*, *might*, and *could*.)

decimate This word means reduce by one-tenth. If you mean, destroy or reduce to nothing, use the words *destroy* or *obliterate*.

disinterested This word means unbiased and impartial. If you mean, bored or having no interest, use the word *uninterested*.

irregardless There's no such word. Use *regardless*.

less Use *less* for uncountable quantities such as *less time* or *less water*. Use fewer for countable quantities: *fewer* birds, *fewer* pages, *fewer fans at the game*.

media The plural form of the singular, *medium*. Don't write, *The media is trying to improve his chances of becoming our next Prime Minister*, as if the word *media* represents a single entity.

real This adjective should not be used as an adverb, as in he's *real* intelligent. Use the adverb *really* to modify another adjective.

reason is because Use "*The reason is that* I've become very interested in grammar."

since, as, for the reason that

Since, *as*, and *for the reason that* are often substituted for *because*, especially at the beginning of sentences. *For the reason that* is wordy and awkward, so it should not be used. *Since* and *as* are poor substitutes for *because*: each word has other, much more prominent meanings. *Since* is a preposition and conjunctive adverb that helps state time relationships (*since* I left college; *since* Thursday afternoon). *As* operates as a preposition, conjunction, or pronoun to express comparisons or processes (*as* you know; *as* he walked along the beach; *as* far *as* I can see; *as* a matter if fact; such provinces *as* Alberta). **Here's a sensible suggestion: whenever you mean *because*, write *because*, no matter where it appears in the sentence.**

Capitalization

Capitals appear at the beginning of these word groups:

- **Complete sentences:**

  In English, all sentences begin with a capital letter.

- **Formal statement following a colon:**

  She tries to instil a strong work ethic in her children, by often declaiming her favourite saying: *If* it's worth doing, it's worth overdoing.

- **Complete sentence quoted within a sentence:**

  He retorted with a stern command: *Do* not even think about leaving this office until you have completed the entire report.

- **Proper nouns and adjectives:** For a full list of capitalized proper nouns and adjectives, consult a grammar handbook such as the *Simon & Schuster Handbook for Writers*. Here's a sampling of capitalized words and phrases:

Queen Elizabeth II	*Grammar and Style at Work*	Member of Parliament	Fred McMurray
Norwegian skiers	Eastern Canada	the Maritimes	Hockey Canada
Selkirk Mountains	planet Earth	Hospital for Sick Kids	McDonald's
Acura RDX	Physics 111	Humboldt Broncos	Passover
World War III	the Quiet Revolution	the Koran	NAFTA
MS Word	Wood Products Division	Acer computers (**but** Acer Inc.)	

Note:

Some words pertaining to the World Wide Web (which has retained its status as a proper noun along with its capitalization) seem to have lost their proper noun status. Examples include *internet* and *web* (when used in phrases such as *internet search*, *web presence* and *webpage*). The term *Web site* seems to have morphed into *website*. However, *the Web* retains the capital letter when it acts as an abbreviated version of *the World Wide Web*.

Abbreviations

Although abbreviations often appear in tables, lists, charts and forms, you should use them sparingly in sentences. Here are a few common abbreviations:

- **Titles:**

  Mr. (Mister) Mrs. **or** Ms. (Mistress) Dr. (Doctor) Miller Ltd. (Limited)

- **Initials used as names:**

  RCMP NATO UNICEF PVR RRSP

- **Names of provinces and states:**

  Postal code designations for provinces and states use Canada Post's abbreviations (BC, AB, SK, WA, and so on). Note, however, that the Canadian branch of the IEEE prefers standard abbreviations (B.C., Alta., Sask., Wash., and so on) within documents. Better yet, write out the name of the province or state.

Numbers

IEEE Canada's rule of thumb for numbers is to write out the numbers one through nine, and to use Arabic numerals for 10 and above. That rule works well for any technical document.

Here are some other usage hints.

- When a number appears at the beginning of a sentence, write out the number:

  Forty-four cartons of 18-year-old single malt whiskey have disappeared from our warehouse.

- Use numerals for numbers one through nine if they're part of a list with larger numbers:

  The results pleased the instructor: The final grades included 4 marks of 90 percent or better, 12 marks from 80 percent to 89 percent, and 14 marks from 63 percent to 79 percent.

- Use numerals for percentages. Write *percent* in normal text, but use % in tables, forms, and charts.

  Sales have jumped 45.5 percent compared to last June's figures.

- When two numbers appear next to each other in a sentence, spell out the first number and use an Arabic numeral for the second:

  We've installed fourteen 8-foot rolls of C350 turf reinforcement mat.

Some style guides and business organizations prefer to use a numeral for the figure that is larger or more difficult to spell:

> Along the Corman Spillway's banks, Melrose Environmental has installed 274 eight-foot rolls of C350 turf reinforcement mat.

- Use numerals for dates, page numbers, ages, times with a.m. or p.m. designations, monetary figures, mileage figures, and addresses (but write out the names of streets numbered less than 100):

  29 March 2009 page 3 12.35 a.m. $3.09 123 kilometres 1245 Monroe Avenue 78A Forty-fifth Street

- Use numerals for decimals and precise measurements:

  456.99556.1 milligrams 87.16 kilos 16 ¾

- Use decimal points for dollar amounts only if cents are included:

  $23.78 $24 $1.48 million $328,098 $16,453.98

Proofreading Hints

Errors are not necessarily easy to detect, even when you know how to spell the words you're using or you know how to use commas effectively. Grammatical errors and unclear phrasing are particularly hard to identify when we've just finished phrasing a document. The following hints try to compensate, then, for our natural tendencies to miss certain kinds of errors, inconsistencies, and poor phrasing.

1. Delay the revision/editing/proofreading process as long as possible, so that you can return to the document with a fresh perspective. Even a 10-minute break will help you detect errors that you will not see at the moment you complete a draft.
2. Proofread more than once.
3. Have others proofread your material.
4. Proofread and revise content and structure first. Then, look at grammar, sentence structure, punctuation, and mechanics.
5. Isolate the spelling and other mechanics from context. (For example, read backwards or highlight random sections.)
6. If you have a tendency to misspell (or mis-key) a certain word, use your word processor's search function to find every placement of that word in the document. Also, use that search function to locate every time the word *this* appears; then ensure that every placement of *this* is followed by a noun or noun phrase. (In Microsoft Word 2010, click on the *Home* tab, then click on *Find* at the far right. Then, key in the word or phrase you wish to find, and every instance of that word or phrase in the document will be listed on the left and highlighted in the document. It's magic!)
7. Highlight paragraphs or other short sections on a computer screen.
8. Don't trust automated proofreading software, which for example wouldn't notice if you've used *loose* instead of *lose*.
9. How to detect errors:
    - Use a grammar handbook to check all aspects of grammar, punctuation, and mechanics that you are not certain you know well.
    - Use a thesaurus and dictionary to find exactly the right word for each meaning.
    - Use lists of commonly misspelled words, commonly confused words, and commonly misused words (pages 78–80 in this workbook) to help with word choices and spelling.

Practice session—mechanics

The following sentences may contain errors in grammar, punctuation, spelling, capitalization, abbreviations, and numbers usage. Underline each error and, in the space provided, correct the error. If a sentence contains no errors, write ***C*** in the space provided. You will find answers for these exercises on page 132.

1. Great chef's can prepare Low Calorie meals that taste evrey bit as good as meals loaded with calories.
2. 14 students came late to the exam they where denied entrance to the exam hall.
3. Boy Im glad I didnt have to study latin in school.
4. The concensus amoung my friends is that we should take off the 2009–2010 year so that we can live and work at the whistler resort during the 2010 olympic games.

5. I went to see a councillor about my grades, that were to low. Theyt said that I should just work harder.

6. Whole areas of Thailands beach front tourist resorts and locals homes were wiped out by the soonamee in December, 2004.

7. The Thai government reported eitght thousand four hundred and fifty-seven confirmed deaths in the wake of the soonamee caused by the 2004 Indian Ocean earthquake on 26 December, 2004.

8. More than three years after the massive earthquake and subsequent soonamees on 26 December, 2004, that devastated countries around the Indian Ocean, recovery programmes undertaken by Red Cross and it's partner national societies continue to provide assistance to those communities most affected by the disaster.

9. Canadians were particulualy generous in providing aid to the countries affected by that disaster.

10. The R.C.M.P. face criticism for several resent insidents which have eroded trust in our national police force.

Improvement exercises—mechanics

The following sentences may contain errors in grammar, punctuation, spelling, capitalization, abbreviations, and numbers usage. Underline each error and, in the space provided, correct the error. If a sentence contains no errors, write ***C*** in the space provided. You will find answers for these exercises on pages 132 and 133.

1. This mp3 player has over three Mb of memory it can store thousands of songs.
2. 46% of all hockey games that go to Overtime are settled in the first 5 minutes of the 1st Overtime period.
3. the Acura mdx offers a three-hundred Horsepower motor.
4. The office has ordered 3 Dell Inspiron one five-two-five lap tops for field work.
5. Dell offer twenty-four technical support fore all its' products.
6. Although dell is the most prominent of the On-line computer sales company's there are several other similar outfits in operating in Canada.
7. The client responded with a very negative assessment of our proposed approach: this proposal fails to meet even our most basic requirements.
8. Since the Royal Banke Cup was inaugurated in 1996, two teams have only won the cup twice; the Aurora Tigers (2007 and 2004) and the Vernon Vipers (1996 and 1999).

9. The North American Free Trade Agreement (nafta) has supporters and detractors in all 3 countries Canada, the United States and Mexico.
10. Dr Judy diSantos, MD is a leading expert on the affects of workplace secondhand smoke.
11. He is a prime candidate for our quality control position, as he had similar experience in the steel industry.
12. The structure was completely decimated when the roof beams gave way under the snow load.
13. They should of reported the cracks that were developing in the support beams.
14. The 3 main principals of successful road construction are illustrated by the Howth project.
15. Always insure that the lock down lever has been properly set at the three o'clock position.
16. in Canada there are four hundred and thirty-four post-seconadry colleges and universities accredited by provincial governments.
17. Beside the public funded and accredited institutions, 49 for profit institutions have been authorized to grant specific credentials.
18. If you know wear to look youll find all kinds of scholarships and burrsories available to Canadian and foreign students too.
19. The AUCC [association of universities and colleges of Canada] administer more than 150 scholarships, fellowship and internship programs.
20. Canadian's standard of living has long depended on its competitiveness in a International trading economy where our natural resources afford an advantage
21. Insuring that postsecondary educational opportunities are affordable and that no qualified individual is denied access because of financial circumstances.
22. For the reason because of greater competition perhaps editing rates are considerably less in the United Staes than in Canada.
23. Universities have received roughly $11 billion in federal research funding since 1997, most of that flowing through research granting councils.
24. In 2006 The University of Toronto enrolled over sixty-four thousand full time students.
25. The highest 2007–2008 college and university tuition fees were at BCs Trinity Western University at $16500.
26. York University's course handout kits are classified carbon neutral, that means they are produced using environmentally responsible printing practices that include incorporating locally produced papers manufactured using sound forestry practices and increased recycled fibre content
27. A carbon footprint measures the impact our activities have on the environment in terms of the amount of greenhouse gases we produce, measured in carbon dioxide units.
28. Virtually anybody can make biodiesel, as it is so easy you can even make it in your own kitchen.

Document critique—mechanics

The following draft document may contain errors in grammar, capitalization, punctuation, abbreviations, numbers, or spelling. Underline all errors and identify them using standard proofreading marks. (See the Correction Symbols list on page 144.) You will find answers for this exercise on page 134.

Note: the following propaganda piece responds to material found at the *Carbon Footprint* website, http://www.carbonfootprint.com/index.html. Assume that the following draft is being prepared to discredit *Carbon Footprint* and other players in the movement that seeks to convince Canadians that climate change is caused by human activity.

Disclaimer: The opinions in this sample document are not necessarily those of the author or publisher of this grammar handbook.

Lets examine some of the claims made by environmental opportinists like Carbon Footprint, a company that makes money from selling carbon offsets.

Claim number one:

> "At Carbon Footprint, our belief is that climate change will only be addressed if each and every one of us takes responsible steps to **REDUCE** their CO_2 emissions."

I've been looking for evidence that CO_2 emmissions one, actually cause global warming and that two, global warming is necessarily a bad thing. Here's what I've found at their website: I see scientific evidence that temperatures have been rising (point seven degrees celcuis in the past 100 years, world wide) and that glaciers have been melting (twenty-thousand square kilometres of Arctic ice in 30 years). But I don't see any evidence that released CO_2 has had anything to do with it. Yes, the increase in CO_2 emissions has been linked with warming, with evidence of those patterns going back four hundred thousand years but where's the connection? One has to prove **correlation**; not just timing. I mean you could just as easily say that periodic global warming cause greater CO_2 emmissions and that interspersed global cooling caused lower CO_2 emmissions.

Claim Number two:

"Even though the hole in the ozone layer doesn't contribute to global warming. We should still be concerned about the affects of increased UV light."

Irregardless of their claims Carbon Footprint hasn't proved this either. Why should we just take their word for it? The company doesn't provide no evidence at their website that the "hole in the ozone layer doesn't contribute to global warming". None, Zilch.

The important thing to remember about Carbon Footprint:

"From this web site only - you can offset your CO2 emissions while you shop on-line. Every purchase you make on-line via the links on this page will count towards our carbon offset programme - including reforestation and clean energy projects globally."

So, we see what Carbon Fooprints all about – money! When they make you feel guilty enough, then you buy their services and you shop with their partners, like eBay and Amazon books and napster. Or buy their Ultra bright windup Eco Torch.

Don't be fooled by what the media gives you; don't be fooled by outfits Carbon Footprint in the business of selling guilt about global warming. Let's put are energy into dealing with the impacts of warming instead of sucumbing to the arrogant human asumption that we can cause a whole planet to change it's climate.

VOCABULARY AND STYLE

The headline in a news story reads "Mad Students Target Teacher." The story is about nine grade-three students who hatched a plot to kidnap and possibly kill their teacher, all because the teacher scolded one of them for standing on a chair. Were they indeed "mad" (i.e., insane) or just angry? Legally, the distinction is extremely important, of course. Possibly the headline writer deliberately used the slang version of *mad* to appeal to readers, or perhaps the headline writer does not know the differences between *mad* and *angry*.

In technical writing, word choice can dramatically affect the reader's perception of the message. Key words also provide critical information, so writers should choose their words carefully. At the same time, technical writers must write concise, focused prose: wordy, flamboyant phrasing does not impress the readers of proposals, reports, manuals, and business correspondence.

Choosing the Right Words

Let's be clear about why people read business and technical writing. Workplace readers seek pertinent information, logical analysis, and sensible conclusions and recommendations. That's all. They don't expect entertainment or in-depth philosophy.

> **Grammar at Work: clear communication is critical**
>
> "When I was a student, I remember my instructors telling us about the importance of concise and accurate communication. Despite this good advice, I didn't attempt to develop good oral or writing skills, always letting interest in other areas take priority. During my working years I continued to see this same mistake in many others, particularly in technically educated people who in other ways were very capable.... I still get frustrated occasionally when attempting to convey ideas and concepts to others. However, clear, persuasive phrasing has been absolutely critical in my career."
>
> —***Dennis Willoughby****, P.E., retired Director of Operations for IPSCO's Tubular Division, and a Consultant who advises re: equipment design and product flow for new pipe mills*

Choosing the right words is essential to meeting workplace readers' needs. Here are several factors that should govern your vocabulary choices:

- **Choose the right technical level** for your reader or range of readers.

  The 6th Canadian edition of *Technical Communication* illustrates three levels of technicality, highly technical, semi-technical, and non-technical. The following example lists three descriptions of an automotive tire, Nokian's Hakkapeliitta WR "All-Weather Plus" product.

  - *Non-technical level:* This introduction to a consumer review is aimed at consumers.

  The Nokian WR is a contradiction in terms: this all-season tire carries Transport Canada's severe snow rating, the only all-season tire to achieve that rating. Even many winter tires don't have it.

  - *Semi-technical level:* This expert review is aimed at knowledgeable consumers.

The WR features a silica rubber compound and an "arrowhead" directional tread pattern that is designed to expel water and snow while allowing the tread blocks to remain in contact with the ground. Its "driving safety indicator" displays numbers in the tire's tread, to gauge the tire's remaining life.

- *Highly technical level:* This description is aimed at technically trained tire reps and wholesalers.

The WR All-Weather PLUS is made of an HD silica compound that offers excellent grip. Lateral grip numbers vary from 0.82g to 0.875g, which rank the WR in the 91st percentile and higher. Its hysteresis rolling resistance (HRR) has been calculated by 3D finite element modeling to be less than 68% of the norm. The tire features an "arrowhead" directional tread pattern that expels water and snow while the tread blocks continue to contact the ground. A unique feature called the "driving safety indicator" incorporates numbers displayed along the centre of the tire's tread in addition to the usual wear bars. A new tire's numbers read "8 6 4" indicating the number of millimetres of tread depth remaining. As the tire wears and the higher numbers disappear, drivers can gauge how much life is left in their tires. A proprietary information pin provides temperature readings at road height. The tire carries the Severe Service Emblem. Sizes range from 235/75R15 (6.5 rim) to 275/40R20XL (9.5 in. in rim). Median tire weight is 32.5 lbs. and tread depth is 14/32.

- **Use specific, concrete words.**

General words name broad classes of things such as *tire*, *computer*, and *job*. Often, such general terms need to be clarified by more specific terms.

General Word	Specific, Concrete Words
tire	Nokian WR G2
computer	Dell Inspiron 15R laptop
job	Senior Technical Writer, Resource Energy Solutions, Calgary

- **Use precise, accurate phrasing.**

Don't write "the reset button *on the dash*" when your reader needs to know the exact location of that button (as in "the reset button *3.4 cm. to the right of the ignition slot*"). Here are some other examples of vague or abstract phrasing that needs to be more precise:

Abstract Term	Precise Version
a luxurious condominium	the condominium has imported tile, glass walls, large rooms with views of the lake, and a built-in audio system
the area has an abundance of old growth trees	the south-facing slopes of the area are covered primarily by Douglas Fir (Cascade dry cool variant); 90 percent of the observed specimens stand 100' or more.
the walkway is a disaster	several of the support pillars have been washed away and over half the planking has rotted to the point that it will not support loads over 100 pounds
the computers are outdated, and they need replacement	The computers have only 1 MB of RAM and 100 GB hard drives. We require a minimum of 3 GB Dual Channel RAM, 300 GB hard drives, and 768MB graphics cards.

An imprecise phrase can also offend a reader. If, for example, someone's report misses a 9 a.m. deadline, do you respond that the report is *late*, *tardy*, *behind schedule*, *overdue*, or *missing in action*? The wrong choice could be disastrous.

Or an ambiguous phrase may simply confuse a reader, as in *send us more sturdy gauges*. Do you want sturdier gauges, or do you want more of the sturdy gauges?

- **Be careful with "tricky" words.**

Certain words have created problems for business and technical writers. For example, use the following extreme words very carefully to avoid misunderstandings and lawsuits: *final*, *all*, *complete*, *best*, *maximize*, *minimize*, *optimize*, *full*, *none*, *any*.

Also, avoid or define these multiple-meaning words: *run*, *top*, *get*, *inspect*, *supervise*, *control*.

"Promise" words such as *guarantee*, *warrant*, *certify*, *ensure*, *assure*, *insure* can also lead to problems. DPIC's loss prevention handbook, *Lessons in Professional Liability*, a publication provided by insurers for architects, engineering firms, and other consultants, recommends not using promise words, or else carefully defining them in proposals, contracts, and progress reports.

The DPIC handbook also advises that if you write *will*, you should be prepared to fulfill the contractual promise implied by that word. If you're not confident that you can meet that obligation, then use *may* or *attempt to* instead of *will*.

Here's a list of some "tricky" words:

	Possible Meanings
shall, will	indicate positive affirmative action—a promise (in specifications, "shall" refers to what the writer requires; "will" is used to express what the writer promises, and under what conditions)
minimize	reduce to a minimum, to the least possible amount or degree
inspect	examine in detail, observe, observe and record all observations in detail, apply
supervise	superintend, oversee, look after (work or workers)
control	the power or authority to manage, direct, restrict, regulate, superintend, govern, administer, or oversee ability or power to command
guarantee, warrant, certify	promises the named act will definitely happen **and** the guarantor will pay if the promise is not kept
furnish	equip with what is needed
install	set in position and connect or adjust for use
provide	to supply; to make available
approve	to sanction, consent to, confirm (**not** conditional permission)
satisfactory	work generally conforms to identified criteria

- **Use analogy to sharpen the meaning.**

  Ordinary *comparison* shows similarities between two things *of the same class* (two computer keyboards, two methods of cleaning dioxin-contaminated sites). *Analogy* shows some essential similarity between *two things of different classes* (report writing and computer programming, computer memory and post office boxes).

  Analogies are good for emphasizing a point ("Some rain is now as acidic as vinegar"). They are especially useful in translating something abstract, complex, or unfamiliar, as long as the more common subject used in the analogy is broadly familiar to the audience. Analogy, therefore, calls for a particularly careful analysis of the audience.

  Analogies can save words and convey vivid images. The following sentence from a description of a trout feeder mechanism uses an analogy to clarify the positional relationship between two working parts:

  > The metal rod is inserted (and centred, **crosslike**) between the inner and outer sections of the clip.

  Without the analogy, "crosslike," we would need something like this to visualize the relationship:

  > The metal rod is inserted, **perpendicular to the long plane and parallel to the flat plane**, between the inner and outer sections of the clip.

  This second version is doubly inefficient: more words are needed to communicate, and more work is needed to understand the meaning.

  Besides naming things vividly, analogies help explain things. The following analogy from the Congressional Research Report helps us understand something unfamiliar (dangerous levels of a toxic chemical) by comparing it to something more familiar (human hair).

  > A dioxin concentration of 500 parts per trillion is lethal to guinea pigs. One part per trillion is roughly equal to the thickness of a human hair compared to the distance across North America

- **Choose words that convey a suitable tone.**

  The tone of your writing reflects your personality and your attitude. The tone depends on (1) the distance you impose between yourself and the reader and (2) the attitude you show toward the subject. Assume that a friend is going to take over a job you've held. You've decided to write your friend instructions for parts of the job. Here is your first sentence:

  > Now that you've arrived in the glamorous world of office work, put on your track shoes; this is no ordinary manager-trainee job.

  First, we notice that the sentence imposes little distance between the writer and the reader (it uses the direct address, "you," and the humorous suggestion to "put on your track shoes"). The ironic use of glamorous suggests that the writer means just the opposite: that the job holds little glamour.

  For a different reader (the recipient of a company training manual, for example), the writer would have chosen some other opening:

  > As a manager trainee at GlobalTech, you will work for many managers. You will spend little of your day seated at your desk.

The tone now is serious, no longer intimate, and the writer expresses no distinct attitude toward the job. For yet another audience (those who will read an annual report for clients or investors), the writer again might alter the tone:

Manager trainees at GlobalTech are responsible for duties that extend far beyond desk work.

Here, the businesslike shift from second- to third-person address makes the tone too impersonal for any document addressed to the trainees themselves.

> **Grammar at Work: proofread for tone**
>
> "Tone and timing are critical in my documents. Proposals, progress reports, invoices, collection letters—they all have to be sent at exactly the right time. And they must have the right tone. Of course, we proofread for errors and for technical accuracy, but we also check the tone, which must be respectful and businesslike. Every once in a while, I have to use a firm tone with a client who wants to cut corners or is slow to pay, but even then I don't want a judgmental or angry tone to creep in...."
>
> —***Ken Langedyk***, *consulting engineer*

Follow these guidelines:

- Use a formal or semi-formal tone in writing for superiors, professionals, or academics (depending on what you think the reader expects).
- Use a semi-formal or informal tone in writing for colleagues and subordinates (depending on how close you feel to your reader).
- Use an informal tone when you want your writing to be conversational, or when you want it to sound like a person talking.
- Above all, find out what the preferences are in your organization.

♦ **Maintain an objective tone.**

You can write objective prose partly through identifying your physical point of view:

The downflow meter is positioned in the pipe, underneath the bridge deck. We recorded its position and current condition with a video camera that we manoeuvred into position at the northeast end of the pipe. Sufficient light was provided by using the camera's built-in LED light.

Also, maintain objectivity in your phrasing:

Don't write Clearly the crew is woefully understaffed.

Write The five-person crew is four days behind schedule, even though they have worked 10-hour days; three additional workers are required.

Using third person helps achieve an objective tone. The first person seems natural and sensible for reporting direct observations or actions (as in "we installed battery-powered smoke alarms in each hallway of the four-storey residence"). However, first person is not always necessary and may even intrude, as in the following example, where it would be better to focus on the recommendations, not on who is recommending the actions:

In this report, *we have outlined* all the necessary measures that Glendale Developments will need to follow to develop this site, including the design of a storm water system.

In brief, *our recommendations according to land development guidelines are as follows*:

Here's a more objective version, which sounds more logical and less prone to personal opinion:

This report outlines all the necessary measures the Glendale Developments will need to follow to develop this site, including the design of a storm water system.

In brief, *land development guidelines have led to the following recommendations:*

- **Use simple, familiar language.**

In school, you may have been rewarded for using the biggest words you could find, but business and technical readers won't necessarily be impressed.

Grammar at Work: use the right word, not the biggest word

"... intelligent people use their large vocabulary only to give clear, exact meaning—never to show off. Big minds use little words; little minds use big words."

—*From "Clear Writing: Ten Principles of Clear Statement" © the Gunning-Mueller Clear Writing Institute*

If two words express essentially the same meaning, always choose the word with fewer syllables. (Write *use* instead of *utilize*, or *mislead* instead of *prevaricate*.)

Remember that *so* means the same as *accordingly*, that *start* or *begin* could replace *activate*, and that *get* means the same as *acquire*.

We're talking about **plain language**, the rallying cry for a worldwide movement that tries to make government documents, legal statements, scientific descriptions, business disclosures, and technical reports easy to read and understand.

Grammar at Work: why use plain language?

"Plain language results in end products that meet readers' needs. Plain language is a courtesy to your readers, telling them what they want to know quickly and clearly.

"Plain language doesn't just happen. It is planned. Planning includes understanding the subject, determining what you want to achieve with the document, and knowing who all the readers are, inside and outside your organization."

—*Judith Whitehead, Vankellers Editorial and Writing Services*

The Practice Exercises on page 103 provide you with a chance to choose plain-language alternatives for pompous phrasing. If you're interested in learning more about plain language, visit these online resources:

- The U.S. Securities and Exchange Commission has placed its Plain English Handbook at www.sec.gov/pdf/handbook.pdf (highly recommended!)

- British writers have led the way in the fight against pompous, garbled writing: www.plainenglish.co.uk.
- In the U.S., Bill Clinton and Al Gore promoted plain English. Their efforts have led to this initiative: www.plainlanguage.gov.

♦ **Avoid over-reliance on jargon.**

Every line of study and work has its specialized vocabulary, its jargon. Such terms save time and promote understanding, as long as the terms are understood by the reader, not just by the writer. Even then, jargon can be overused and can get downright silly. For example, a Canadian engineer invented the term, *modular vertical retaining device*, which he promptly reduced to the acronym *MVRD*, for obvious reasons. Can you guess what he meant in phrases such as "the MVRD in the southwest quadrant"? He meant "the southwest wall"! That's right; he thought "wall" was too simple a word for an educated person to use.

On the other hand, a jargon term like SAIL (the Stanford Artificial Intelligence Lab) has a very specific meaning for people in artificial intelligence research. These days, you can keep up specialized vocabularies through such sites as *Cool Jargon of the Day* (where I found "SAIL") or Jon Storm's list of computer jargon at www.jonstorm.com/glossary/.

♦ Please **avoid useless jargon** of the type immortalized in Philip Broughton's buzzword system. Examples include *systematized third-generation hardware* and *balanced incremental programming.* We can't reproduce his system here, but you can find it at these sites:

http://nathancreative.com/buzzphrase.html

www.acronymfinder.com/buzzgen.asp

www.copywriting-on-demand.com/buzzword-generator.htm

Honeywell has taken the systematic buzzword projector a step further. The Simplified Modular Prose Writing System (SIMP) is described at www.copywriting-on-demand.com/THINK-SIMP-LY.pdf.

Writing Concisely

Busy workplace readers appreciate concise, readable documents that don't waste their time. This workbook demonstrates on pages 47 and 48 that **short sentences** make reading easier. Now we're going to see how to trim unnecessary words within sentences.

♦ **Avoid unnecessary words in phrases.**

Many phrases can be simplified with no loss in meaning:

Wordy	Concise
an innumerable number of tiny veins	countless tiny veins
at this point in time	now
bright green in color	green (or, bright green)
absolutely complete	complete
during the time that	When (or, during)

completely free	free

See the Practice Exercises on page 103 for more phrases that you can trim.

- **Avoid unnecessary qualifiers.**

Wordy	Concise
personally, I believe	I believe
very unique	unique
owing to the fact that	because

- **Use active verbs instead of linking verbs.**

Wordy	Concise
I am of the opinion that	I believe
There are not many people who can present a report on this subject in so precise a manner.	Few people can report this subject so precisely.

- **Avoid *there is*, *there are*, and *it is* constructions.**

Wordy	Concise
There is a short in the cable that can interfere with signal splitting.	A short in the cable can interfere with signal splitting.
There are several reasons why we cannot hire this candidate.	Several reasons prevent our hiring this candidate.
It is necessary for them to change the study method.	They must change the study method.

- **Place action in verbs; axe extra verbs.**

In the following examples the real action lies buried in a noun or adjective.

Wordy	Concise
Give me a call.	Call me.
Provide a recommendation for …	Recommend …
Ralph *is* a man who enjoys a task that is challenging.	Ralph enjoys a challenging task.
The technologist *has* the job of translating the engineer's designs into working plans. [Here, the action is placed in the object.]	Technologists translate engineering designs into working plans.

- **Avoid unnecessary prepositions.**

Wordy	Concise
under the impression that …	believing that …
at the present time	now
in the event that	if

The editor *of* the company newsletter recently received an award *for* writing.	The company newsletter editor recently received a writing award.

Grammar at Work: don't waste words

When writing a business plan,

- Avoid unnecessary jargon
- Economize on words
- Use short crisp sentences and bullet points
- Concentrate on relevant and significant issues
- Break the text into numbered paragraphs, sections etc.

—From PlanWare's Plan Write Business Planner website, at http://www.business-planware.com/business-plan-when.htm

Writing Clearly

This workbook's "Basic Grammar" section illustrates techniques that make connections clear:

- Make each pronoun refer to one easily-identified antecedent (page 17).
- Ensure that each pronoun agrees with its antecedent (page 18).
- Always write a noun or noun phrase immediately after a reference pronoun such as *this*, *that*, *these*, or *those* (page 17).
- Use the right pronoun case (page 15).
- Make subjects and verbs agree (page 25).

The workbook's "Sentence Grammar" chapter shows how to create clear, readable sentences, using techniques such as the following:

- Write straightforward SVO and SVC sentences (page 47).
- Keep subjects and verbs together (page 50).
- Avoid overlong sentence openers (page 50).
- Avoid faulty coordination (page 53).
- Avoid faulty parallelism (page 54).

Grammar at Work: leaders must communicate clearly

"A little mist in the mind of a leader becomes dense fog in the minds of followers."

—Don Gibbs, Chief Production Officer, tekmar Control Systems

Finally, the "Punctuation" chapter shows how commas and other punctuation are so important in conveying your exact meaning.

You should also **avoid ambiguous phrasing** such as

> Most students will party on Friday night.

Does the above sentence mean that the students will go to a party this Friday night, or do they "party" every Friday night? And who are "most students"? Most of the students at a particular college or university, or students in general? And would those students be at Canadian schools, or is the writer referring to all North American students?

Here are some possible meanings that the writer had in mind:

> Most students whom I know party every Friday night.
>
> Most of the students in my dorm are going to Friday night's party at Nevin Hall.
>
> Most Canadian university students relax at Friday night parties.

Yes, the context in which the sentence appears could probably help a reader get some sense of the writer's intended meaning, but why should the reader be forced to engage in detective work? Ambiguity works well in poetry—it helps enrich the bounty of possible meanings that bounce off each other—but ambiguity is a definite no-no in technical writing. Consider these statements:

> All steel is prone to rust. [Under what conditions? For example, does steel rebar rust when it's encased in concrete? And does "all steel" include stainless steel?]
>
> Set up the monitoring station at the corner of 8th Street and Monroe. [That intersection has four possible setup locations—SE, SW, NE, and NW.]

- **Avoid overstuffed sentences.**

  A sentence crammed with ideas makes details hard to remember and relationships hard to identify, as in the following example.

Overstuffed	Revised
Publicizing the records of a private meeting that took place three weeks ago to reveal the identity of a manager who criticized our company's promotion policy would be unethical	In a private meeting three weeks ago, a manager criticized our company's policy on promotion. It would be unethical to reveal the manager's identity by publicizing the records of that meeting.
The basic subject (*Publicizing the records*) is too far from the verb (*would be*).	Other versions are possible here, depending on the writer's intended meaning.

So give your readers no more information than they can retain in one sentence.

- **Un-stack modifying nouns.**

  One noun can modify another noun (as in "software development"). But when two or more nouns modify a noun, the string of densely packed words becomes hard to read and ambiguous.

Stacked	Revised
Be sure to leave enough time for a **training session participant** evaluation. (*Evaluation of the session or of the participants?*)	Be sure to leave enough time **for** participants **to evaluate** the training session. *or,* Be sure to leave enough time **to evaluate** participants **in the** training session.

In the following sentence, in which adjectives are stacked in front of a noun, readers can readily see that the adjectives modify *employee*:

He is a temperamental, independent, aloof, but brilliant employee.

- **Avoid ambiguous modifiers.**

  Modifiers explain, define, or add detail to other words or ideas. If a modifier is too far from the words it modifies, the message can be ambiguous.

The plant foreman delivered a rousing speech at the meeting, <u>an advocate of flexible work hours</u>.

In the above sentence, it's hard to tell that the foreman is *an advocate of flexible work hours* because that modifying phrase is closer to *meeting*. Instead, the sentence should read:

<u>An advocate of flexible work hours</u>, the plant foreman delivered a rousing speech at the meeting.

The limiting modifier *only* is often misplaced. Here's one example:

The Act <u>only</u> allows engineers to approve the designs.

When *only* is placed before *allows*, it modifies that word when the writer meant to say:

The Act allows <u>only</u> engineers to approve the designs.

Other examples:

Most of my classes are online, so I <u>only</u> go to the campus on Thursdays.

Smoking is <u>only</u> allowed in the gazebos on this campus.

Write readable paragraphs

Most of the paragraphs we write are standard body paragraphs that provide information, description, and analysis. Technical subjects can become rather dense with detail, so clear paragraph development is essential. Here are four critical aspects of readable paragraphs:

1. Each paragraph must deal with only one subject or a specific aspect of a subject.
2. Each paragraph requires a topic statement that identifies the paragraph's topic and signals the purpose of the paragraph. That topic statement might require one sentence, two sentences, or perhaps just part of a sentence. (Pages 178–179 in the 6th Canadian edition of *Technical Communication* discuss topic statements.)

3. Each paragraph should use the pattern that best suits its content and purpose. Examples of patterns that suit technical topics appear on pages 181–186 in *Technical Communication.*
4. Transitional words and phrases link ideas within a paragraph.

   In the following paragraph, the topic statement is underlined and the transitions are italicized.

   For the Text Enrichment Site (TES) that will accompany the 6th Canadian edition of Technical Communication, I propose a new "Job Search" section that includes some old and some new features. *Previously posted material* includes an application package, *but* that package *now* forms the basis for an exercise that asks students to evaluate the applicant's letter and résumé. *Other new material* would include a list of online resources for job seekers, a sample "compliance matrix" for a specific job search, and a hyperlinked electronic job search portfolio that includes streaming video. *I'm particularly excited about the electronic portfolio* because no other TES that I've seen has this kind of business writing sample.

Now here's a paragraph that doesn't really have an effective topic statement, partly because it tries to cover two topics. It also lacks transitions and it seems to wander off topic.

> Smoke can contain several types of particles. Fires exposed to abundant oxygen burn at high temperature and with small amount of smoke produced. Smoke particles might be mostly composed of ash, or even condensed aerosol of water. High temperatures produce nitrogen oxides. Sulfur content yields sulfur dioxide. Carbon and hydrogen are almost completely oxidized to carbon dioxide and water. Fires burning with lack of oxygen can produce many compounds. These are often toxic. When carbon doesn't completely burn, it produces carbon monoxide. Nitrogen-containing materials can yield hydrogen cyanide, ammonia, and nitrogen oxides. Fires are often dangerous to fight.

The following edited version now includes two paragraphs, each with its own topic. Each paragraph has a topic statement (underlined) that says what the paragraph is about. Each now includes useful transitions (in italics) that link everything together.

> *The makeup of smoke depends on the fuel and on the available oxygen*. Fires exposed to abundant oxygen burn at high temperatures and with small amounts of smoke produced: smoke particles are mostly composed of ash, or even condensed aerosol of water. High temperatures also produce nitrogen oxides and sulfur content yields sulfur dioxide. In those hot fires, carbon and hydrogen are almost completely oxidized to carbon dioxide and water.
>
> Fires burning with lack of oxygen can produce many compounds, many of them toxic. *For example*, when carbon doesn't completely burn, it produces carbon monoxide. *Also*, nitrogen-containing materials can yield hydrogen cyanide, ammonia, and nitrogen oxides. *Therefore*, these types of fires are often dangerous to fight.

Here are some transitional words and phrases:

- **go-ahead words**—and, therefore, furthermore, also, for instance
- **new-idea words**—thus, so, and so, therefore, consequently
- **summary words**—as a result, at last, finally, in conclusion, as we've seen
- **change-idea words**—but, yet, nevertheless, otherwise, although, despite, however, conversely, on the other hand
- **link-cause-and-effect words**—that caused, as a result, that produced, consequently
- **referring words**—they, these, though, not one, all but two, without exception

- **restricting and qualifying words**—provided, in case, if, lest, when, occasionally, even if, never

Writing Strong Prose

Clear, concise writing helps convey your credibility to the reader, as do the technical content and the overall structure, and the visual presentation of your document. Now, here are some other factors that help convey that sense of authority needed for your reader to take your document seriously.

- **Use the active voice.**

  Don't write "The report was written by me."

  Write "I wrote the report," which is shorter and more powerful.

  The difference between active and passive voice can be more dramatically shown in a longer passage, such as these geotechnical recommendations concerning the foundation design of a proposed hockey arena. (Italics identify the passive verbs.)

  Passive voice, wordy and annoying because it's not easy to read:

  As indicated in Section 4.1, *it is recommended* that the entire building *be founded* on a minimum 1.0 meter thick layer of granular fill placed and compacted as specified in Section 4.2. If the *above is carried out,* then the proposed arena *can be supported* by spread and/or strip footings and *designed* using an allowable bearing pressure of 145 kpa.

  For the heavier column loads, *it is recommended* that these piles *are supported* with piles end bearing on bedrock. *It is anticipated* that the piles could range from about 2.0 meters to in excess of 15 meters within the center and west side of the proposed building location, respectively. However*, it is recommended* that additional boreholes *be drilled* to confirm the actual pile length.

  [124 words]

  Active voice, more concise and more readable:

  As Section 4.1 indicates, we recommend founding the entire building on at least 1.0 meter of granular fill, placed and compacted as specified in Section 4.2. Then, spread and/or strip footings can support the proposed arena, if designed for an allowable bearing pressure of 145 kpa.

  We recommend supporting the heavier column with piles that bear on bedrock. These piles could range from about 2.0 meters in the center of proposed building location to more than 15 meters at west side. However, additional boreholes *should be drilled* to confirm the actual pile length.

  [93 words]

  In the above edited version, underlined words indicate the active voice. Notice that the edited version retains one case of passive voice, to emphasize what is done rather than who should drill the holes. (Who drills the holes is irrelevant at this point.)

- **Reserve the passive voice for special meanings.**

  Although you should write in the active voice whenever possible, you'll find the passive voice useful when you want to emphasize what happened, rather than who made it happen, or if you do not want to offend the reader:

  The project has been completed on time, and on budget.

  The remediation work has not been completed according to the terms of the contract.

- **Provide specific facts and figures** when they're required to support an argument or to impress the reader.

  Don't write This proposal will benefit the City of Johnsonville's water system.

  Write Implementing the water treatment methods described in this proposal will remove the threat of a Cryptosporidium outbreak in the City of Johnsonville's water system.

  Don't write A high water table caused the clay to swell and create a large broken mound in the basement floor.

  Write When the water table rose to within 4.5 meters of the surface, the clay became swollen, resulting in a roughly rectangular mound 7.4 meters long, 1.73 meters wide, and .82 meters high in the center of the basement floor. The center of this mound is cracked along its long axis. (See Figure 2.)

- **Use direct phrasing.**

  Readers appreciate sentences that get to the point as quickly as possible. Yes, many sentences need to begin with a transitional word or phrase, but the following sentences open with superfluous or awkward phrases:

  Indirect prepositional phrase and subsequent passive voice:

  From the previous assessments completed (Section 3.3.1), potential bottlenecks were identified, and in some cases, assigned improvement priorities. [18 words]

  More direct:

  The previous assessments (Section 3.3.1) identified potential bottlenecks and, in some cases, assigned improvement priorities. [15 words]

  Indirect prepositional phrase and subsequent passive voice:

  With the methods described above, and having an inspector on site during critical production periods, damage will be minimized. [19 words]

  More direct:

  These methods, monitored by an on-site inspector during critical production periods, will minimize damage. [14 words]

- **Fight noun addiction.**

  Verbs should carry the action in a sentence. Weak writing, however, often buries the action in nouns manufactured from verbs. To make matters worse, this kind of phrasing has to add prepositions to create recognizably English patterns.

Weak and Wordy	**Strong and Concise**
We ask for the **cooperation** of all employees.	We ask that all employees **cooperate**.
provide **recommendations** with regard to	**recommend how to**
conduct an **investigation** of	**investigate**
effect construction of	**construct (or, build)**
Monitoring of the complete deactivation works was done in 2000.	**The complete deactivation works were monitored in 2000.**

manage **establishment** of	**establish**
do an **analysis** of	**analyze**
has the capability of **monitoring**	**can monitor**
make a discovery of	**discover**

Besides causing wordiness, a nominalization can be vague—by hiding the agent of an action.

Weak and Vague	**Precise**
A valid requirement for immediate action exists. (*Who should take the action? We can't tell.*)	We **must act** immediately.

- **Address readers directly when appropriate.**

Formal reports and proposals do not lend themselves to personal pronouns such as *I* and *you*. Although it is natural, perhaps, to write "we conclude," it is better to write "this report's findings suggest" or "the evidence points to." Exceptions might occur when (1) you need to specify that you performed a task ("I tested the brake system three times") or when (2) you wish to emphasize a report recommendation ("We recommend four solutions"). Still, that recommendation could be phrased "This report has shown that four solutions are required."

However, in letters and emails you should use *you* and *your* to connect with readers. An **impersonal tone** results from

Students at our university will find the faculty always willing to help.

Readers will react more favourably to the **more personal tone** in

As a student at our university, *you* will find the faculty always willing to help.

Practice session 1—choose vocabulary that suits the reader's technical level

You will find answers for these exercises on page 134.

Which of the following instructions is appropriate for the first-time user of an Android tablet? Why is the most appropriate of the three instructions?

1. Turn the device on.
2. To turn the device on, press and hold the **Power/Lock Key**.
3. Pressing the **Power/Lock Key** will energize the Quad Core, 1400 MHz processor.

Practice session 2—use precise, accurate phrasing

Revise the following sentences to make them more precise, accurate, or informative. You will find answers for these exercises on page 135.

1. Our outlet does more business than Montreal.
2. This car has a very high top speed.
3. Confusion is in control of this office.
4. Your crew damaged a piece of office equipment.
5. We had a lot of rain last night.

Practice session 3—use an objective tone

Revise the following sentences to make them more precise, accurate, or informative. You will find answers for these exercises on page 135.

1. The group's spokesman demanded overdue salary increases.
2. There's absolutely no justification for granting your request for a promotion.
3. As we have clearly shown in this report, the contractor is at fault for using sub-par materials in building the bridge. [Found in the conclusion of a 110-page formal report].
4. Every act has a consequence, as shown by the theft of company equipment following your shocking breach of security.
5. I can't approve your irresponsible proposal to increase our marketing budget by 20 percent.

Practice session 4—use simple, familiar language when appropriate

Revise the following sentences to create straightforward, familiar language. You will find answers for these exercises on page 135.

1. May you find luck and success in all endeavours.
2. I suggest you reduce the number of cigarettes you consume.
3. A good writer is cognizant of how to utilize grammar in a correct fashion.
4. I will endeavour to ascertain the best candidate.

Practice session 5—use jargon sparingly

Revise the following sentences to eliminate ambiguities in phrasing, pronoun reference, or punctuation. You will find answers for these exercises on page 135.

1. For the obtaining of the X-33 printer, our firm will have to accomplish the disbursement of funds to the amount of $3000.
2. To optimize your financial return, prioritize your investment goals.
3. The use of this product engenders a 50 percent repeat consumer encounter.
4. We wish to welcome all new managers aboard.
5. Intercom utilization will be employed to initiate substitute employee operative involvement.

Practice session 6—identify empty jargon and pompous phrasing

Underline all empty jargon and pompous phrases. In the space provided above each line, write simpler or more familiar words or phrases. You will find answers for these exercises on page 136.

I am in agreement with your proposed solution to the overtime problem, which seems increasingly

vexatious to both management and the union, because the current way of allocating overtime

work is unfair to employees and is not in keeping with overall management policy. However, I am not enamoured with the projected costs of your proposed approach, which appears burdensome for our company. Therefore, it is my considered opinion that a re-working of the overtime pay scale will be necessary, particularly because we are in a heavy debt load situation at the moment.

Practice session 7—identify wordy phrasing

Underline wordy phrases and circle unnecessary words in each of the following sentences. You will find answers for these exercises on page 136 and 137.

1. Return that spectrometer to the person named Gordon Burns when you have finished using it.
2. Will the ratepayers make payment of the development costs for the proposed riverside trail system?
3. Some, though not all, air toxins come from absolutely natural sources such as burning forest fires and the wearing away of soil erosion.
4. The Nissan Maxima SE sports sedan automobile that I have for sale is silver in colour.
5. There are several aviation-related businesses that lease land and facilities from YVR.

Practice session 8—reduce wordiness

In the space provided beside each phrase, write a simpler version of the phrase. You will find answers for these exercises on page 137.

fewer in number ______________________

circled around ______________________

for the reason that ______________________

true fact ______________________

goes under the name of ______________________

surround on all sides ______________________

if conditions are such that ______________________

small in size ______________________

in view of the fact that ______________________

I personally feel ______________________

it is often the case that ______________________

repeat again ______________________

proceed ahead ______________________

it would appear that ______________________

final conclusion ______________________________

lenticular in character ______________________________

equally as good ______________________________

oval in shape ______________________________

disappear from view ______________________________

plants exhibited good growth ______________________________

the tube which has a length of 3 m ______________________________

prior to ______________________________

serves the function of being ______________________________

it is possible that the cause is ______________________________

Practice session 9—use active verbs instead of weak verbs

Revise the following wordy and vague phrasing to eliminate weak verbs. You will find answers for these exercises on page 137.

1. Our disposal procedure is in conformity with federal standards.
2. Please make a decision today.
3. We need to have a discussion about the problem.
4. Your conclusion is in agreement with mine.
5. This manual gives instructions to end users.

Practice session 10—write concisely

Revise the following sentences to eliminate needless prepositions and "to be" constructions, and to cure noun addiction. You will find answers for these exercises on page 137 and 138.

1. Igor seems to be ready for a vacation.
2. In the event of system failure, your sounding of the alarm is essential.
3. These are the recommendations of the chairperson of the committee.
4. Our acceptance of the offer is a necessity.
5. Please perform an analysis and make an evaluation of our new system.

Practice session 11—write concisely

Revise the following sentences to eliminate needless phrases, needless repetition, and redundancy. You will find answers for these exercises on page 138.

1. I have admiration for Professor Singh.
2. Due to the fact that we made the lowest bid, we won the contract.
3. On previous occasions we have worked together.
4. This report is the most informative report on the project.

5. She is a person who works hard.

Practice session 12—eliminate needless prefaces

Revise the following sentences to eliminate There and It openers and needless prefaces. You will find answers for these exercises on page 138.

1. It is apparent that there was severe fire damage to the reactor.
2. There are several reasons why Jenna left the company.
3. It is essential that we act immediately.
4. It has been reported by Clayton that several safety violations have occurred.
5. This letter is to inform you that I am pleased to accept your job offer.
6. There are not many people who can present a report on this topic in so precise a manner.
7. There are several key issues that confront our company as we begin this year.

Practice session 13—eliminate needless prepositions and use direct phrasing

Revise the following sentences to eliminate unnecessary prepositions and wordy prepositional phrase openers. Use direct phrasing. You will find answers for these exercises on page 138.

1. Table 5.12 presents the plan for implementation.
2. From the previous assessments completed, potential restoration sites were identified, and in some cases assigned preliminary work priorities.
3. Through preliminary research, we are confident that our plan will benefit the environment.
4. In researching this report, it was useful for learning about project management.
5. With the methods described above, and having a crew on site during critical draw down periods, fish mortality will be minimized.
6. By replacing and modifying key components, and by ignoring the decimal in the display, the speedometer was made to read the wire feed rate in inches per minute.

Practice session 14—eliminate ambiguities

Revise the following sentences to eliminate ambiguities in phrasing, pronoun reference, or punctuation. You will find answers for these exercises on pages 138 and 139.

1. Call me any evening except Tuesday after 7 o'clock.
2. I cannot recommend this candidate too highly.
3. Janice dislikes working with Claire because she's impatient.
4. Despite his efforts, Joe misinterpreted Sam's message.
5. Our division needs more effective writers.
6. Tell the reactor operator to evacuate and sound a general alarm.

Practice session 15—clarify ambiguous modifiers

Revise the following sentences to eliminate ambiguities resulting from misplaced modifiers. You will find answers for these exercises on page 139.

1. The professor did not accept late assignments, determined to maintain his authority.
2. Many people like to relax after a hard day's work in front of the TV.
3. The employee persuaded his boss to grant a pay raise with logical arguments.
4. I want the final Amex report written by your division.
5. He only ordered six laptops for the new office.

Practice session 16—identify paragraphing techniques

Revise the following sentences to eliminate ambiguities resulting from misplaced modifiers. You will find answers for these exercises on pages 139 and 140.

1. Underline the topic statement in the following paragraph.
   - Does that statement cover the paragraph's content? Yes No Partly
   - Does it signal the paragraph's purpose? Yes No Partly
2. If necessary, improve the phrasing of the following paragraph's topic statement, in the space above it.
3. Circle the transitional words in the following paragraph. If necessary, add transitional words in the spaces above the lines.
4. Strike a line (~~like this~~) through any errors in spelling, grammar, or punctuation, and write the corrected version on the line above the error.

You'll like our floors. A quality hardwood floor adds warmth, elegance, and increases the value of your home. It strengthens and stabilizes the inherent structure of your home. Because of ease of maintenance, and hardness, wood flooring allows you to have the healthiest of living environments. Also, hypoallergenic. Our company offers a full range of wood species, colours, sizes, as well as finishes to meet the needs of the most discerning wood flooring buyer. Each floor we deliver reflects the many years of development in state of the art production, our Premium finish is the most wear resistant finish on the market.

Practice session 17—use active voice instead of passive voice

Revise the following sentences to replace weak, wordy passive voice with the stronger active voice. You will find answers for these exercises on page 140.

1. The report was presented by RJ Overton and Bliss Strand, to the board on May 5.
2. Nuisance weeds are considered by our department to be a high priority.

3. The water quality produced by the well is excellent.
4. However, based on our study of local topography and surface water features, it is expected by us that groundwater is moving from east to west.
5. A final report will be prepared by our department upon receipt of your comments.

Practice session 18—use passive voice for special circumstances

Revise the following sentences to use passive voice to emphasize the recipient or what was done, rather than who performed the action. You will find answers for these exercises on page 140.

1. *Edit to emphasize the recipient:* Our project design team estimated the completion dates.
2. *Edit to emphasize what was done:* Channel Construction removed woody debris, boulders, and gravel from the site following a storm in May 2014.
3. *Edit to emphasize what happened rather than who made it happen:* In the 2013 Texas hold 'em Grand Finale, Harvey Grace won $1.7 million on the very last hand with a Royal Flush after frequently bluffing his opponents in earlier deals.
4. *Edit to emphasize what is being provided:* We present recommendations for mitigating potential impacts of the bridge construction.
5. *Edit to emphasize the result:* The proposed work will not create improved access to the beach.

Practice session 19—place the action in verbs, not in nominalized nouns

Revise the following sentences to use passive voice to emphasize the recipient or what was done, rather than who performed the action. You will find answers for these exercises on pages 140 and 141.

1. Please provide a description of each of the slope failures.
2. An activation of the compression valve will result in rapid filling of the chamber.
3. Once verification is made for fish habitat, we can decide how to design the channel to ensure a minimum of disruption to the spawning grounds.
4. We present recommendations for mitigation of potential impacts of the bridge construction.
5. Please send a response at your earliest convenience.
6. She was unable to give an explanation for the bearing failure.

Improvement exercises—sentences

The following sentences may contain phrasing that is wordy, overrun with jargon, pompous, weak, imprecise, or indirect. If so, improve the phrasing in the spaces provided.

Also, correct errors in sentence structure, pronoun usage, adjectives and adverbs, conjunctions, prepositions, spelling, or punctuation. You will find answers for these exercises on pages 141 and 142.

1. The use of the report would be to enable the Maintenance Department at CanaWood's Kelowna plywood plant to demonstrate feasibility of the roller screw in this application to upper management in order to proceed with in-plant testing.

2. We did research on the ISA server by Microsoft. This is a software firewall that is installed on a computer or server.

3. As you requested, the following progress report is provided with regard to the status of our study on Wastewater Management in Oyama that is being prepared for the Oyama Concerned Citizens Committee.

4. The report will review wastewater management issues for the community and make recommendations for further progress in mitigating nutrient loading to area lakes from wastewater discharges.

5. The following is a summary of the progress to date of our project.

6. The assessment was completed as a requirement in order to secure financing for the project through Natural Resources Canada.

7. Understanding the job requirements and roles in shipping, shift work can be advantageous.

8. In the following memo I will describe the activities of the proposed workshop series.

9. Our research regarding your marketing tools identified a security hole in your external network. This helped us create a software tool that is easy to use, but powerful enough to prevent security breaches.

10. I have often been an attendee at your concerts.

11. The writing instructors are assessing participants' writing competencies and are using their education, experience, available computer resources, and optimum pedagogical practices to provide assistance through on-site counselling to the majority of participants that have dysfunctional barriers to effective communication.

12. In conclusion, with the recent improvements in drainage control along West Sitkum Road, the majority of the recommendations of the 1996 assessment are completed, and the risk of sedimentation to local water domestic water supplies is minimized.

13. With having our computer website, an interactive inventory is required.

14. A few of the plant species were unavailable at the time of planting and were substituted with more of the available species.

15. Within the College there are computer laboratory facilities that offer specialized training opportunities.

16. Water quality monitoring in Richmond Creek upstream and downstream the proposed outfall site should take place, beginning two years in advance of plant start-up.

17. Both species have a preference for colder water as is likely to occur in this lake.

18. Our mortgage software automatically calculates the effects of rising interest rates, your income, various amortization periods, coordinating with fluctuating property values,

property taxes, compares to your pre-approved credit rating, payment schedules, then requests quotes from 18 Canadian lenders.

19. The study area is not within close proximity to the Fraser River.

20. An important component of this initiative will be establishing targets and reporting on achievements against these targets. This will assist in ensuring accountability and transparency in all areas.

Improvement exercise—paragraph

The following paragraph may contain phrasing that is wordy, weak, imprecise, or indirect. If so, improve the phrasing. Also, correct errors in sentence structure, pronoun usage, verb tenses, spelling, or punctuation. Finally, improve the paragraph's structure by creating an effective topic statement and by adding transitions. You will find answers for this exercise on page 142.

> Cork has been used as a flooring material for more than one hundred years (100), it provides outstanding acoustical and thermal insulation, it really cushions the foot, it is resistant to moisture damaga and decay. With its surface treatment of 5 coats of UV-cured acrylic finish which contains no volatile organic compounds, Cork is definitely easy to clean. A light cleaning with a damp mop will be maintaining it's new appearance. Cork flooring is proving to be very durable. Most manufactures offer a good warannty on everything except water damage.

Document critique 1—structure, phrasing, and mechanics

A golf club manufacturer has designed a set of hybrid irons that help "average" players address two failings of "normal" iron sets:

1. Most mid-to-high handicappers (players who shoot in the mid-80s or higher) find that they don't hit their long irons (3 and 4 irons) consistently well and they don't hit the ball far enough with these irons even when they make a good swing.
2. If a golfer buys hybrids—clubs that are easy to hit because they combine the best features of irons and fairway woods—to replace the 3 and 4 irons, then there is a large "gap" between the 4 hybrid and the 5 iron. (A good player with a consistent swing would hit each successive club about 15 yards farther than the previous one; for example, a 6 iron might propel the ball about 160 yards and a 5 iron about 175 yards. However, the gap between a 4 hybrid and a 5 iron might be 25 or even 30 yards.)

Woodsman Golf has designed an integrated set of irons that includes

- two hybrids (equivalents of 3 and 4 irons) and
- nine irons that incorporate technology found in the hybrids.

The irons include the numbers 5, 6, 7, 8, 9, and 10, and the pitching wedge, gap wedge, and sand wedge.

Woodsman's marketing director has written the following draft of a product description that will be "placed" in a golf magazine.

Your task:

Proofread and edit the draft. Look for errors in grammar, spelling, punctuation, and mechanics and improve the phrasing to make it clear, concise, and suitable for an enthusiastic but non-technical audience. Also, improve the paragraphing where necessary. You will find an edited version of this document on page 142.

The Woodsman 993 hybrid set of irons bridges the gap between irons and hybrids. Club designer Ed Ring made those choices. He increased the lofts on all the irons, he replaced the 3 iron and 4 iron with hybrids, the remaining eight irons have hollow heads and extreme perimeter weighting.

The 993 hybrid 3 and hybrid 4. The 3 has 20 degrees of loft, the 4 hybrid has 22. The lofts on the 5, 6, and 7 irons are four degrees stronger than standard, at 24, 27, and 30°. The 34° 8 iron is 3 degrees stronger, the 9 iron is two degrees stronger, at 39°. Then the pitching wedge is 44° as compared to the standard 48°, the gap wedge is 50°, the sand wedge 55 degrees.

Hollow heads, a thin face, a huge cavity back is filled with a light (only 5 grams) carbon composite cap. The 35 grams thus saved have been moved to the bottom and back of the head. The Iron has a low deep center of gravity that launches the ball high. Combining this high launch angel with less loft than normal makes the ball go further with each club. A person who would hit a ball 135 yards with a standard 8 iron will hit it 145 or 150 yards with the woodsman 993 7 iron. That's important because the ball will still come in nice and high with good spin, so it will stick on the green.

At the top end of the set, an average player will hit the ball 175 yards with the 5 iron and 190 yards with the 4 hybrid a useful gap between clubs. The set is available with Rifle steel shafts or with Aldila graphite shafts. Designed to match the 993 heads.

Note: GolfSpyder.com explains "loft" at www.golfspyder.com/golf-club-loft.html

Notes and Answer: document critique

Document critique 2—structure and phrasing

A student has drafted the following summary of a report that has resulted from a geotechnical investigation of the subsurface and bedrock conditions that underlie the proposed site of a hockey arena. Although some details of the summary have been fictionalized, the report is based on an actual situation.

Here's your task:

1. Read the entire summary to understand what it's about. If necessary, consult a dictionary or online search engine to learn the meanings of terms such as *surficial* and *berm.*
2. Review and, if necessary, edit each paragraph in the document to ensure that each has an effective topic statement, that the paragraph flows logically and smoothly, and that the reader can easily move from one paragraph to the next.
3. Improve the sentence structure and phrasing where necessary.
4. Correct errors in grammar, spelling, punctuation, and mechanics.

You will find an edited version of this document on page 143.

SUMMARY

With the proposed project, the purpose of the investigatin was to determine the subsurface soil/bedrock conditions at the proposed building location and based on our interpretation of this information, to provide comments and recommendations thereby pertaining to the geotechnical aspects of the proposed arena site, locating on the actual northwest corner of the intersection of seventeenth avenue and Corcoran road.

Therefore the results of the investigation indicates that the east half of the proposed building site consists of a bedrock knoll while the west is underlain by 0.5 to 4.0 meters of fill materials followed by up to 0.9 meters of loose organic soils. The fills and/or organic soils are underlain by native deposits of stiff clayey silt and clay. The bedrock surface which will require drilling and blasting to the proposed design grade elevations and native stiff clayey silt and clay will provide a suitable base on which the proposed arena facility. As indicated in Section 4.1, it is receommended that the entire building be founded on a minimum 1.0 meter thick layer of granular fill placed and compacted as specified in Section 4.2. A soil weight of 2000 kilograms per cubic meter may be used in the design calculations. Observations and borehole results indicate that approximately ¾ of the proposed arena site is underlain by end dumped fills ranging from 0.3 to 1.8 meters thick followed by a thin layer of surficial organic soils. The above surficial soils and fills are underlain by an interlayer sequence of clay, silty clay, and clayey silt. It is also understood that the proposed design will include an approximate 3.7 meter high berm around ¾ of the proposed building. The influence of this around the proposed arena, could result in an additional 50 to 70 millimeters of recompression settlements. In order to minimize potential total and differential srettlements of the proposed structure it is recommended that consideration be given to a preloading of the proposed area site in question.

A alternat option for locating the actual location of the arena considers the proposed areana to be located within the north portion of the site where the fill thickness is relatively thin and/or nonexistent. The granular fill that presently overlie the proposed arena site have essentially performed as a preload that has been in place for an estimated period of about twenty-one years, more or less. It is therefore logically concluded that prefoundation treatment of this site would only involve over excavating to remove any further fills and/or surface organic soils beneath the proposed building. That would be cheaper than the option proposed and investigated in this report's report.

Smith Consulting Ltd. has carried out a geotechnical investigatin at the above noted site. To confirm the inferred surface conditions testing, included moisture content determination and Atterberg Limit tests.

[465 words]

Notes and Answer: document critique

ANSWER KEY

Diagnostic Test

1.	E	16.	E	31.	B	46.	A
2.	D	17.	D	32.	A	47.	A
3.	E	18.	C	33.	A	48.	E
4.	C	19.	A	34.	B	49.	B
5.	E	20.	E	35.	B	50.	E
6.	A	21.	A	36.	B	51.	A
7.	E	22.	B	37.	B	52.	A
8.	B	23.	B	38.	A	53.	B
9.	D	24.	A	39.	B	54.	A
10.	C	25.	B	40.	A	55.	A
11.	C	26.	B	41.	D	56.	B
12.	B	27.	A	42.	B	57.	A
13.	A	28.	B	43.	D	58.	B
14.	D	29.	B	44.	D	59.	B
15.	B	30.	A	45.	E	60.	A

Nouns

Practice session—nouns

1. specialties
2. Celsius
3. shelves
4. editor-in-chief's
5. RVs and boats
6. Ryan Collins's
7. pit boss
8. Bill's and Wanda's reports
9. data, datas
10. 1980s

Improvement exercises—nouns

In items 1 through 5, underline the common nouns and circle the proper nouns.

1. Each of the consultants wrote a report about Riverbend Dam's projected cost overruns.
2. In this book, "grammatical" refers to effective, well-formed phrases and sentences.
3. Grammar studies the rules governing the use of language, as Dr. Benson has stated many times.
4. Written technical communication is not the only subject covered in *Technical Communication*, a book designed for present and future technical communicators.
5. Brick walls can withstand considerably higher temperatures than concrete board walls.

In items 6 through 10, underline the subjects and circle the objects.

6. Each of the consultants wrote a report about Riverbend Dam's projected cost overruns.
7. In this book, the term "grammatical" refers to effective, well-formed phrases and sentences.
8. Grammar studies the rules governing the use of language, as Dr. Benson has stated many times.
9. Written technical communication is not the only subject covered in *Technical Communication*, a book designed for present and future technical communicators.
10. Brick walls can withstand considerably higher temperatures than concrete board walls.
11. truck's
12. criteria
13. Sundin's
14. partners'; company's
15. software
16. procedures than our
17. Rene Corbeil's
18. stimuli
19. phones; signal's
20. shorelines

Document critique—nouns

(The underlined words indicate where errors have been corrected.)

MEMO

To: Harlan Devonshire, Field Supervisor

From: Arjun Atwal, Accounting Office

Date: June 24, 2009

Subject: Expense Claims

We have had difficulty establishing the accuracy of recent expense claims submitted by your exploration group, especially those claims submitted this past month.

Would you please review all your team members' expense claims before they are sent to our office? I'm sure that you know when claims are excessive and you'll be able to correct them before we have to deal with them.

As an example of what I mean, I've attached Rose Francis' claim for this May. I find it hard to believe that she could spend $900 on camp supplies in addition to the food allowance!

ANSWER KEY

Pronouns

Practice session—pronouns

1. me	6. its, their
2. its	7. whom
3. who	8. Toni Colette and I
4. his or her	9. who
5. its, their	10. its

Improvement exercises—pronouns

1. them	6. his or her
2. whom	7. your
3. its	8. We
4. whose	9. me
5. its	10. company's

11. All the team at EBS, including me, look forward to hearing from you.
12. These dense patches of forest are not a vulnerable ecosystem. The forest patches are in generally good condition, but they are found on steep slopes and would be best left undisturbed by development.
13. A new water intake at Lambeau Lake will require a new structure to be built below the current water level of 112 feet. In order to allow construction at an elevation of about 92 feet, the lake level has to be dropped at least 20 feet.
14. The housing used to protect the stream meters is a 2-inch galvanized pipe, which rests flush with the riprap boulders on the left bank.
15. The system includes two cigar-sized dataloggers that contain both pressure transducers and temperature recorders. These dataloggers are housed in protective casings.
16. Canadian National Railways (CN) has used a GIS-based railway Information System to map the location of all known streams that are crossed by its railways.
17. Box type concrete culverts were constructed with a concrete skirt (Figure 15) downstream of the outlet, in order to reduce scour that would block fish passage below the skirt.
18. Certain specifications govern the transfer vinyl records to MP3 files.
19. Just between you and me, our network specifications will not meet Microsoft's standards.
20. Martin Technologies has just announced who it has chosen as its main supplier of cable and peripherals.
21. <u>It</u>'s time for the company to review all <u>its</u> recruiting practices.
22. Frances and Donna have each told their clients about the new line of excavators.
23. Austin St. Pierre of St. John's Ambulance Services will explain its new billing policy.

24. In Calgary, officials ended their (intersection) red light camera program after statistics showed a 40 percent drop in ticketable offenses.
25. Researching this report taught me about management techniques.

Document critique—pronouns

(The underlined words indicate where errors have been corrected.)

Concept Architectural Designs

MEMO

To: Devin Marchand, Architectural Technologist

From: Shirley Tostenson, Managing Partner

Date: April 16, 2009

Subject: Preliminary Research for Potential project

We are in the preliminary stages of preparing a proposal to design Halifax's potential Seaport Redevelopment project. I am assigning you to conduct research that will place our proposal in context.

Please review all the available literature on the following group of Waterfront developments: Granville Island, Vancouver; the Distillery district, Toronto; The Custard Factory, Birmingham, England; The Forks, Winnipeg. Make note of each development's capitalization, scope, end uses, community involvement, sources of capital, and architectural themes, and its blend of commercial, retail, arts and crafts, and residential uses.

Complete the survey by April 25 and bring the results directly to me. This is a high priority, so drop everything else that you are doing. Don't discuss this outside of the office; we don't want this leaking out to those who might be also submitting a proposal.

Verbs

Practice session—verbs

1. is	6. needs
2. have	7. his or her.
3. sings	8. was, were
4. was	9. were
5. was	10. laid

Improvement exercises—verbs

1. Our group conducted the experiment.
2. We'll prepare a final report when we receive your comments.
3. *A list of upcoming squash games has been posted on the bulletin board.* Passive voice suits this sentence because it's irrelevant who posted the list.

4. Following a storm in March 2008, woody debris, boulders, and gravel were removed from the site.
5. The engineering firm submitted its proposal three days early.
6. The following report summarizes the project's progress to date.
7. The report also includes comments that estimate ecosystem richness and the presence of invasive alien species.
8. Successive approximations converts analogue to digital signals, by comparing the unknown input against a known input produced by the D/A converter.
9. Our firm uses their software for tracking traffic on our website.
10. Freeware.com provides the best option for reliable data on who uses open source software.
11. In March 2009, Northern Nurseries and Landscaping began its reconstruction of MacDonald Creek's banks.
12. A new crib wall was built in July 2008.
13. I believe that online gambling threatens small local casinos.
14. The test method and the answer key are faulty.
15. The system includes two cigar-sized dataloggers, each of which contains a pressure transducer and a temperature recorder. The dataloggers are housed in protective casings.
16. My instructor's facilitative role provides learning opportunities.
17. I will be graduate from George Brown College's Health Informatics program on May 26.
18. No work program was planned for at the time of this environmental assessment.
19. Allowing the engine to breathe easier typically produces 15 more horsepower.
20. The following analysis combines an assessment, a comparison, and a recommendation for doubling next year's research budget.
21. Every time I suggested an alternative plan, my boss raised budget objections.
22. The guidance document's rough draft recommends labeling pharmaceuticals that would pose an environmental risk when released into the water supply.
23. He must have grown four inches over the summer holidays!
24. The Accident Prevention Board likes to hold informal sessions that include all stakeholders.
25. The Accident Prevention Board comprises seven members from a wide range of industrial occupations.
26. There are some oranges in the fruit bowl.
27. Each of the winners receives $500 and his or her choice of Corel software.
28. The media have their own agenda regarding the Kyoto Accord.

Document critique—verbs

(The underlined words indicate where errors have been corrected.)

Concept Architectural Designs

MEMO

To: Shirley Tostenson, Managing Partner
From: Devin Marchand, Architectural Technologist
Date: April 22, 2009
Subject: Preliminary Research for Potential project

As you have requested, I have researched several waterfront developments. As you will see in the attached research overview, some very exciting developments have happened in Canada and in England.

I have found most of the information about the respective water front developments from the internet. However, most of the internet material is supplied by the development corporations themselves. So, that information is suspect. That's why I then went to newspaper archives for the Vancouver, Toronto, Birmingham, and Winnipeg daily papers. The business sections were especially useful.

Also, I was able to get documents from the planning offices in the city of Vancouver and the city of Winnipeg. I have a friend in City Hall in Toronto; she was able to give me background details of the proposals submitted to the City for the Distillery District development. I received those details last week. I have phoned the City Clerk in Birmingham. He promised to send a package of materials that describe the whole Custard Factory development, complete with construction costs. I hope the package comes because there was not much else available.

Thanks for assigning me this research project, which has been very interesting.

Adjectives and Adverbs

Practice session—identify adjectives and adverbs

1. Two dedicated teams desperately want to win the Vanier Cup, which is awarded annually to the champion of CIS football.
2. Biodiesel has a higher cetane level than petroleum diesel.
3. The Canadian cement industry claims that it has steadily improved energy efficiency in its manufacturing processes.
4. Improved indoor air quality probably reduces the number of sick days claimed by workers.
5. Despite urgent requests from the local health authorities, the mine repeatedly refused to install modern water treatment methods.
6. He has done well with the aging equipment at his disposal.
7. The Earth Home system provides fully integrated, self-sufficient buildings.
8. Compressed earth bricks are environmentally sustainable because they require comparatively little energy to produce and very little energy to transport.
9. The sales manager will work closely with the plant manager to plan just-in-time deliveries.

10. Someone has inadvertently left the specimen case on the loading dock.

Improvement exercises—adjectives and adverbs

1. Don't think badly of me because I tried to improve our working conditions.
2. According to Bobbie Cruz in the IT department, every computer hard drive eventually fails.
3. They advised us to move slowly and carefully when working with those volatile chemicals.
4. We must conserve depleted, scarce, non-renewable resources or face the consequences.
5. Reported rollover rates for vans and SUVs are considerably higher than for sedans.
6. He was clearly the stronger of the two of them.
7. The maple syrup harvest is way earlier than usual this year, perhaps because of global warming.
8. The researchers made an honest mistake when they miscalculated the size of the honey bee population in Ontario.
9. Global cooling surely is more likely than global warming in the next 50 years.
10. ***C***
11. In March 2009, Northern Landscaping discovered a near collapse of the Arduin Reservoir dike. *Adjective:* He ran near the prison walls.
12. He is usually late for meetings that require him to present a project update. *Adjective:* The train arrived two hours late at Victoria Station.
13. Please report daily about the progress of labour contract negotiations. *Adverb:* The *Globe and Mail* delivers its daily paper to my door.
14. Three times this season he has been reprimanded for playing fast and loose with the fishery regulations. *Adverb:* The new BMW is a very fast car.
15. She responded to the dissident workers' grievances very well, don't you agree? *Adverb:* I am not well.
16. Please submit the monthly production figures by the second day of the following month. *Adjective:* We get paid monthly.
17. Do you think that our boss is a well woman? *Adjective:* Trial lawyers need to speak well in public.
18. The early indications are that the real estate bubble is about to burst. *Adjective:* If you don't arrive early, the best seats will be taken.
19. Just how hard did you try to make the exam questions difficult? *Adverb:* Yes, that was a hard exam.
20. *The Globe and Mail*'s weekly "Career" section has recently focused on the changing demographics of the Canadian work force. *Adjective:* I check the water heater weekly because we've had so many problems with leaking heaters.

Document critique—adjectives and adverbs

(The underlined words indicate where errors have been corrected.)

Christopher Metals

MEMO

To: Shift Foremen

From: Augusta Brown, General Manager

Date: May 22, 2009

Subject: Reducing Toxic Emissions At Our Plant

As you have might have heard, we received a D⁻ from Environment Canada following their December site visit. Apparently we have 6 months to clean up our act – the inspectors will be here again in November. If our sulphur and hydrochloric acid readings have not been reduced by 50 percent, they'll move to close us down.

To be honest, I don't think we've been doing that bad. For example, our culvert mill and our lead coating facility don't have any more free-flowing smoke stacks – the emission capture equipment diverts the gases to the holding chamber where the gases are chilled and the toxic elements are filtered out. But that's not good enough for the government, I guess.

I'm attaching some literature on equipment that's supposed to reduce sulphuric and hydrochloric acid emissions. Would you please read the descriptions and look at the diagrams of the equipment to see whether we could fit such devices into our existing setups? I mean, you know your areas inside out, so you'll know if the equipment will work in our plant.

Pay particular attention to the rugged low-cost low-maintenance air pollution control equipment sold by Herring Engineering. How much of it would we need? Can we cut into our ductwork to install it? The price looks right, so look at the Herring stuff really carefully.

Let's discuss this next Tuesday in our regular production meeting.

Conjunctions, Prepositions, and Interjections

Practice session 1—identify prepositions

1. The Vanier Cup is awarded annually [to the champion] [of CIS football].
2. Improved indoor air quality probably reduces the number of sick days claimed [by workers].
3. Despite urgent requests [from] the local health authorities, the mine repeatedly refused to install modern water treatment methods.
4. The news media are very interested [in the global warming issue].
5. Someone has inadvertently left the specimen case [on the loading dock].

Practice session 2—identify conjunctions

Conjunction	Type of Conjunction	Relationship Expressed
and	coordinating	connects the compound subject, "research and understanding"
that	subordinating	introduces subordinate clause, "longer autumns may be ..."
but	subordinating	introduces subordinate clause, "they release CO_2 when ..."
so	subordinating	introduces subordinate clause, "more CO_2 is being absorbed ..."
however	conjunctive adverb	acts as transition into contrasting statement
so	subordinating	introduces subordinate clause, "more CO_2 is being released ..."
because	subordinating	introduces subordinate clause, "soil microbes are ..."
and	coordinating	connects two subordinate clauses

Practice session 3—identify prepositional functions

Prepositional Phrase	Type of Modifying Phrase	Word or Phrase Modified
in Ottawa	adverb	says where the owl "captured media attention"
in the House of Commons	adverb	says where the petition was tabled
for a national park	adjective	modifies "plans"
in the South Okanagan	adjective	modifies "park"
by B.C. Southern MP Alex Atamanenko	adverb	says who submitted the petition
to a meeting	adverb	says when
of the National Park Network	adjective	says what kind of meeting
with representatives	adjective	says more about the meeting—with whom?
with Parks Canada officials	adverb	modifies "worked"
of view	adjective	modifies "points"

Improvement exercises—conjunctions, prepositions, and interjections

1. Where is the new computer projection <u>system?</u>
2. Clearly, we have to improve the communication <u>among</u> the production team, the sales people and the shipping people.

3. Either the transmission bolt threads are stripping when the engine torque increases because they're made of too soft steel <u>or the engine might be too</u> powerful for this type of transmission.
4. Michelle damaged the portable oscilloscope when she tried to jam it <u>in</u>to a metal container that wasn't large enough.
5. Please remember to not place the GPS receiver <u>beside</u> equipment that emits a strong magnetic field.
6. The maple syrup harvest <u>involves way more</u> work than most people realize.
7. Way to <u>go! That</u> set of schematics represents the best work you've done for our firm.
8. Big hybrid SUVs, <u>such as</u> the Toyota Highlander, the Chevrolet Tahoe, and the GMC Yukon, are now getting up to 30 miles per Imperial gallon.
9. Neither the diehard Ottawa football fans <u>nor</u> the corporate community believe that a CFL football team can flourish in Ottawa.
10. ***C***

Document critique—prepositions and conjunctions

(The underlined words indicate where errors have been corrected.)

Christopher Metals

MEMO

To: Shift Foremen

From: Augusta Brown, General Manager

Date: June 2, 2009

Subject: Reducing Toxic <u>Emissions </u>at Our Plant

Thank you for <u>your</u> useful <u>comments</u> in <u>yesterday's</u> meeting. I know we differ with each other on some points, but overall <u>our</u> experience <u>points</u> <u>toward</u> three <u>alternatives</u>.

Alternative 1 – upgrade <u>our</u> emission capture equipment. I thought that <u>our </u>existing equipment is state of the-art<u>; however</u>, as Jim reported, <u>it's</u> already showing signs of aging and needs new filtering and storage <u>add-ons</u>. This alternative is probably the least expensive.

Alternative 2 – shut down the galvanizing facility. This section of the plant requires the most expensive work<u>, but</u> it generates 65 percent of <u>our</u> income. So I don't want to shut it down.

Alternative 3 – shut down the culvert mill, sell its equipment, and use the sale funds to expand and modernize <u>our</u> galvanizing facility. Not only would this strategy get rid of the culvert mill, <u>which</u> is <u>barely covering</u> operating <u>costs, but</u> it would allow us to expand production of galvanized pipe, <u>which</u> is in high demand. Beside these benefits, <u>as Jeanine pointed out</u>, new ways of capturing and storing galvanizing by-products have been <u>developed</u>.

<u>There</u> seems to be more research to do, as we look <u>into</u> this issue. <u>Let's</u> meet again on Monday.

ANSWER KEY

Sentence Exercises

Practice session 1—identify subjects and predicates

1. You will need to return that spectrometer to Gordon Burns.
2. Will the ratepayers pay the development costs for the proposed riverside trail system?
3. Some air toxins come from natural sources such as forest fires and soil erosion.
4. Air toxins pose a serious threat to human health in all parts of the world.
5. The ozone layer in southern Canada thinned by an annual average of about 6 percent in the late 1990s.

Practice session 2—identify sentence types

6. complex
7. simple
8. compound-complex
9. compound-complex
10. simple

Practice session 3—identify sentence errors

Underline each example of a comma splice (CS), fused sentence (FS), dangling modifier (DM), misplaced modifier (MM), sentence fragment (frag), disruptive modifier (DisM), faulty coordination (coord), or faulty parallelism (//) in the following sentences. In the space above the underlined error, write the appropriate symbol (CS, DM, etc.)

11. The Ekati mine has cost Dia Met about $536 million and (**coord**) is 70% complete.
12. When grading the site (**DM**), two-way radios, which allow surveyors to communicate from a distance, are used.
13. Most people like to relax after a hard day's work in front of the TV (**MM**).
14. After completing the project under budget (**DM**), the head engineer rewarded me with a promotion.
15. The employee persuaded his boss to grant a pay raise with logical arguments (**MM**).
16. Many municipalities created, for the benefit of local farmers (**DisM**), policy that provided for minimal monitoring of animal waste runoff.
17. The Integration comparator is also quite popular, (**CS**) it is used most commonly in digital multimeters.
18. When manufacturing a car chassis (**DM**), materials need to be strong and lightweight.
19. The objectives of this assessment were to ascertain whether the completed work is in compliance with the prescriptions and that (//) the drainage along the road might have been improved.

20. The lake will be temporarily drained during the drawdown (**FS**) it will then increase in size during the re-wetting phase of the project.

Improvement exercises—sentences

1. Storm water pond construction, which was completed following best management practices, did not cause negative environmental impacts.
2. The plan area, which includes about 390 ha, is bound by Powers Creek to the east, Highway 97 to the north, the District of Peachland to the west, and Okanagan Lake to the south.
3. If the native trees are lost or damaged during the construction process, we'll plant three trees for every one that needs replacing. This over-planting will ensure an 80 percent survival of shrubs and a 100 percent survival of trees after three years.
4. Watering these plants will ensure quick root growth if they're watered twice weekly for the first two weeks and depending on the weather, if they're watered once a week for four weeks thereafter.
5. The entire ESB team looks forward to hearing from you.
6. One of his finer projects was the Penticton Channel recreation area; he helped design this project and he oversaw its construction.
7. This proposal outlines the proposed research that will need to be conducted and the solution to the problem that we will implement.
8. Erecting a breakwater wall will divert the currents, thereby eliminating beach erosion.
9. Facial expressions include raising or lowering the eyebrows, squinting, clenching the jaw, and many others.
10. pH measures the acidity or alkalinity of water. Many organisms that live in the water are sensitive to changes in pH.
11. These steps will help protect you from the sun's harmful rays: keep sun exposure to a minimum, especially between 10 a.m. and 3 p.m., wear wide-brimmed hats, wear sun screen with Sun Protection Factor of 15 or greater, and reapply the sun screen frequently.
12. Google Inc. tried to persuade members of the public to refrain from using "google" as a verb because it feared the dilution of its trademark. Google Inc. remembered what happened to brand names like Kleenex and aspirin and trampoline, and Google noted the increasing popularity of the word, to the point where the word became entrenched in dictionaries such as the Oxford English Dictionary.
13. On the RAR database, we will submit a post-development/construction report that summarizes the development, outlining the degree of compliance with the above measures and demonstrating that the SPEA is protected.
14. The proposed "F"-shaped dock, which consists of a 1.8 m wide mainstem, will be made of perforated material and.
15. ***C***
16. The average person spends 3,220 hours stuck in traffic, but only minutes planning his or her retirement.

ANSWER KEY

17. Client A has not attended regularly, but he continues to make contact with our counselors.
18. Smoking is allowed only in the gazebos at this campus.
19. Each of us has a responsibility to the organization.
20. We observed the clients at the Centre, trying their very best in the aptitude tests.
21. If you know the answer, it is easy. However, not everyone knows the answer.
22. After you send the message, the following screen will appear.
23. Specialty NiCd batteries have a niche market in the area of cordless and wireless telephones, emergency lighting, model airplanes, and power tools.
24. With a relatively low internal resistance, a NiCd battery can supply high surge currents. This feature makes them a favourable choice for remote controlled electric model aeroplanes, boats, and cars, as well as cordless power tools and camera flash units.
25. This workshop series has three objectives:
    - developing strategies for structuring documents,
    - learning how to proofread and edit other's documents, and
    - improving ones' efficiency in writing.
31. Please feel free to contact me if you require further clarification.
32. This first tier of the assessment process is completed by the supplier.
33. Roughly 100 pharmaceuticals have now being identified in rivers, lakes, and coastal waters throughout Europe and in north America, in concentrations that have alarmed environmental protection agencies.
34. The European Medicines Agency recommends additional testing of drugs to learn how they break down when they enter surface or underground water supplies.
35. One worry for pharmaceutical companies is that increased amount of testing could translate into costly delays for the release of new drugs.

Document critique—sentences

(The underlined words indicate where errors have been corrected.)

Corrine Maltais

From: Billie Cochrane <bcochrane@on.gc.ca

To: Corrine Maltais <cmaltais@on.gc.ca

Sent: Friday, June 27, 2008 11.19 AM

Subject: Conflicting Responsibilities for the Georgian Bay Project

As my project leader, will you answer my question about what I'm supposed to be doing on the boat. When I took on this summer student placement, I thought that I and the other student would just be observing procedures and collecting data, not doing all the grunt work on the boat. Isn't that what sailors are for?

Here's what been happening –

- On Monday, I and Jeannie (she's the other student) had to coil all the ropes and check for flaws in the ropes. That's very hard work, let me tell you: those ropes are heavy and some of them are dirty, too.
- Then on Tuesday the captain made us clean the decks. He said that we're the ones getting the deck dirty with all our fish samples and we came on board with muddy boots after checking the mud flats at 3 Moon bay. That's a big boat actually, and we had to use mops and then hand scrub with those hard sponge things. Why don't they have proper motorized cleaning equipment on board?
- Wednesday we sailed to Parker Point to get water samples and to check for fish mortality. Do you remember that we had heard reports that fish were washing up on shore? We didn't see any, though, so maybe the reports were for another point. That was a good day: we collected samples from 17 lakeshore locations and 5 deep water locations, we observed three outflow pipes and we took samples from those outflows. One of them - I think it was the pulp mill around the corner from Parker Point - was really disgusting. The colour was muddy red and it smells like sulphur and diesel fuel, which is a really foul combination for sure.
- As you know, we spent yesterday in the lab with your techs as they checked the samples for toxic substances. They found very high levels of dieldrin and lindan, which was surprising because I thought that those pollutants were in decline. At least, that's what Professor Gorsline said.
- This morning Jeannie and I went down to the boat to retrieve some equipment and the captain said that we had to clean the cabin and the storage lockers down below. "Whoa," I said, "that's not what I came here for. Im a forth year biology student, and Im going to tell Dr Maltais what's going on." And he said, "Go right ahead; I don't care, but its my boat and you'll do what I tell you when you're on board That's what he said. Well, I came right up to see you but you were in a meeting so Jeannie said I had better write to you right away.

Thanks for listening. Will you talk to the captain?

Punctuation Exercises

Practice session—punctuation

1. **a.**
2. **b.**
3. **b.**
4. **c.**
5. **b.**
6. **a.**
7. **a.**
8. **c.**
9. **c.**
10. **c.**

Improvement exercises—punctuation

1. The Trackstick has over 1 Mb of memory, which can store months of travel information.
2. *C*
3. The Trackstick has several applications: employee monitoring, vehicle tracking, speed control, photo tours, public safety, and law enforcement.
4. CEO Hernan Hesse says that the Trackstick II offers “upgraded features and accuracy compared to the original device.
5. The device receives signals from 24 orbiting satellites that precisely calculate its position on this planet.
6. The property is owned by Northern Tent Vacations. The company’s President, Carlo DeValero, has authorized Green Consultants to apply for environmental approvals required for development.
7. Don’t worry about the condition of its transmission; the rest of the car is in really good shape.
8. The proposed wharf will be the same length and appearance and it will use the same materials as the adjacent wharf.
9. She orchestrated and wrote a major update of a maintenance manual.
10. Shortly after he graduated with an engineering diploma, he developed a process for evaluating the quality of concrete as it is poured.
11. Prominent scientists such as Henri Casselle have questioned the popular beliefs about global warming.
12. Art Burns’s designs won the Granby Award for high density housing projects this year.
13. Well, they own the fishing rights, so we should back off.
14. The Canadian dollar—as strong as it is—has hurt U.S. exports that are paid in U.S. dollars.
15. The Leafs post-2005 regular season records leave a lot to be desired.
16. The rocket burst into flames upon re-entry into the earth’s atmosphere.
17. Generation Y workers are not all that different from any other “generation” of workers, according to workplace analyst Barbara Moses.

18. This grammar workbook, which supplements your *Technical Communication* textbook, provides hundreds of exercises.
19. The Spanish say, *hasta mañana* and that's exactly how I feel about wanting tomorrow to be here now!
20. On 10 June 2004, a British Airways Boeing 777 suffered a serious fuel leak as it took off from Heathrow airport.
21. Lack of documentation, or poor documentation, costs your company money.
22. The total cost of a single call answered by a help desk technician is $20 or more, depending on the length of the call.
23. On its website, Innovatia reports on the costs of distributing poorly written documentation.
24. The Regional Aquatics Monitoring Program (RAMP) is an industry funded multi-stakeholder initiative.
25. The Athabasca River has high year-to-year fluctuations in its flow patterns.
26. "I'd be shocked if we learned that the cause of the flooding was one of the valves installed last week," said Roger Desroches, Managing Director.
27. Unconscious bias may cause some doctors to more likely recommend knee replacement surgery to men than to women, according to a recent study.
28. Compared with men, women have the surgery when their arthritis is at a more advanced stage. This fact perhaps explains why women tend to do more poorly than men after knee replacement.
29. The water quality in the Athabasca River in the fall of 2005 was better than in the previous year.
30. On March 25, 2008 the *Globe and Mail* carried a story about Biovail's involvement in a series of frauds that distorted reports of earnings, and revenue.
31. The US Securities and Exchange Commission said Biovail executives were "obsessed with meeting quarterly and annual earnings guidance," and engaged in a "corrupt strategy" to do so.
32. In the United States, Sirius Satellite Radio Inc. has taken over its rival XM Satellite Radio Holdings Inc., which means that the Canadian subsidiaries will also merge.
33. Montreal's waterfront district, once a bustling scene of shipping and manufacturing, has almost become a ghost town. However, current private and public investment plans promise to rejuvenate the area.
34. The Canadian fashion industry is not centered in one location. Rather, it's found in Winnipeg, Montreal, Toronto, and Vancouver, and ,of all places, in Vernon, British Columbia, the home of Far West and Valhalla Pure.
35. Several media networks, including traditional media companies like Forbes Inc., are trying to lure advertisers away from Google Inc. and Yahoo Inc. by offering one-stop shopping for ad placement.
36. Each of the consultants wrote a report about Riverbend Dam's projected cost overruns.

37. In this <u>book, the</u> term "grammatical" refers to <u>effective,</u> <u>well-formed</u> phrases and sentences.

38. Rollover rates for SUVs are nearly twice that of sedans, and 30 percent higher than that of pickup <u>trucks.</u>

39. ***C***

40. <u>Sundin's</u> career as a Toronto Maple Leaf appears to be <u>ending, which</u> is really <u>too</u> bad.

41. Don't expect your cellular <u>phones</u> to receive strong <u>signals</u> in the far northern Arctic.

42. According to Rene <u>Corbeil's</u> <u>report, the</u> bridge is no longer <u>safe,</u> but we knew that.

43. Blogging has real advantages for <u>companie</u>s that possess clear goals and a workforce <u>whose</u> ambitions align with the <u>company's</u>.

44. Guerilla e-mail <u>(www.guerilla-mail.com)</u> supplies disposable e-mail addresses <u>that expire</u> after 15 <u>minutes, so</u> you can register a software program or service but not receive spam later on.

45. A study by York <u>University</u> professor<u> Dr Robert Kozinets</u> shows that consumers of technology are governed partly by their view of the role that technology plays in <u>their</u> lives and in society.

Document critique—punctuation

(The underlined words indicate where errors have been corrected.)

Billie Cochrane

From: Corrine Maltais <cmaltais@on.gc.ca

To: Billie Cochrane <bcochrane@on.gc.ca

Sent: Monday, June 30, 2008 7.48 AM

Subject: Conflicting Responsibilities for The Georgian Bay Project

Thank you for drawing my attention to your experiences on the Great Lake Queen, our research vessel. I have spoken with Captain Blair who's denying the accusation made by you and Jeannie Brock.

Let me respond to each of your comments;

- On Monday, you say that Captain Blair forced you and Jeannie to coil all the ropes and check for flaws. He says that he asked you to help him and the two sailors, but only for 15 minutes because they had a particularly difficult time with the anchor rope that somehow become kinked. He said that after you helped with the anchor rope - which he appreciated by the way - he saw you coiling some dock lines. However, he says he didn't order you to do that.
- He said that on Tuesday all six of you, including him, cleaned the deck.
- On Friday, you went to retrieve the collection equipment. According to Captain Blair, he asked you to clean the cabin because you had left a number of items lying around during Wednesday's trip to Parker Point. Also, he claims that you and Jeannie didn't clean up the galley after you prepared lunch. By the way, the captain said that he and the crew appreciated your offer to cook lunch on Wednesday and it was good, too, he said.

So here's where we're at. I think we (you, Jeannie, the captain, and I) need to get together to discuss your formal roles in the research team while you're on board the Great Lake Queen. We also need to talk about the things that team member do to help each other out and to work as a team. Often, these things go beyond the formal defined roles, you know. I'll set up a meeting for this Wednesday before you sail to The Ramsey shoals. Watch for an email later today, which will set the time we should meet on the boat.

By the way, you mentioned that your Professor had said that dieldrin and lindane are in decline in the Great Lakes. Probably his comments were based on a flawed study conducted by the Michigan Fish and Game League. Their results have since been shown to have been based on a too small sample and solely on levels of those chemicals found in one fish species.

ANSWER KEY

Mechanics Exercises

Practice session—mechanics

1. Great chefs can prepare low calorie meals that taste every bit as good as meals loaded with calories.
2. Fourteen students came late to the exam they were denied entrance to the exam hall.
3. Boy, I'm glad I didn't have to study Latin in school.
4. The consensus among my friends is that we should take off the 2009–2010 year so that we can live and work at the Whistler resort during the 2010 Olympic games.
5. I went to see a counselor about my grades that were too low. They said that I should just work harder.
6. Whole areas of Thailand's beach front tourist resorts and locals' homes were wiped out by the tsunami in December 2004.
7. The Thai government reported 8,457 confirmed deaths in the wake of the tsunami caused by the 2004 Indian Ocean earthquake on 26 December 2004.
8. More than three years after the massive earthquake and subsequent tsunamis on 26 December 2004 devastated countries around the Indian Ocean, recovery programs undertaken by Red Cross and its partner national societies continue to provide assistance to those communities most affected by the disaster.
9. Canadians were particularly generous in providing aid to the countries affected by that disaster.
10. The R.C.M.P. face criticism for several recent incidents which have eroded trust in our national police force.

Improvement exercises—mechanics

1. This MP3 player has over three MB of memory that can store thousands of songs.
2. Forty-six percent of all hockey games that go to overtime are settled in the first five minutes of the first overtime period.
3. The Acura MDX offers a 300 horsepower motor.
4. The office has ordered three Dell Inspiron 1525 lap tops for field work.
5. Dell offers 24 technical support for all its products.
6. Although Dell is the most prominent of the online computer sales companies, there are several other similar outfits in operating in Canada.
7. The client responded with a very negative assessment of our proposed approach: "This proposal fails to meet even our most basic requirements."
8. Since the Royal Bank Cup was inaugurated in 1996, two teams have only won the cup twice, the Aurora Tigers (2007 and 2004), and the Vernon Vipers (1996 and 1999).

9. The North American Free Trade Agreement (NAFTA) has supporters and detractors in all three countries, Canada, the United States, and Mexico.
10. Dr Judy diSantos, MD, is a leading expert on the effects of workplace secondhand smoke.
11. He is a prime candidate for our quality control position, because he has had similar experience in the steel industry.
12. The structure was completely destroyed when the roof beams gave way under the snow load.
13. They should have reported the cracks that were developing in the support beams.
14. The three main principles of successful road construction are illustrated by the Howth project.
15. Always ensure that the lock down lever has been properly set at the 3 o'clock position.
16. In Canada, provincial governments have accredited 434 post-secondary colleges and universities.
17. Besides the public funded and accredited institutions, 49 for-profit institutions have been authorized to grant specific credentials.
18. If you know where to look, you'll find all kinds of scholarships and bursaries available to Canadian and foreign students.
19. The AUCC (association of universities and colleges of Canada) administers more than 150 scholarship, fellowship and internship programs.
20. Canadians' standard of living has long depended on their competitiveness in an international trading economy where our natural resources afford an advantage
21. [We] ensure that postsecondary educational opportunities are affordable and that no qualified individual is denied access because of financial circumstances.
22. Editing rates are considerably lower in the United States than in Canada, perhaps because of greater competition.
23. ***C***
24. In 2006 The University of Toronto enrolled over 64,000 full time students.
25. The highest 2007–2008 college and university tuition fees were at BC's Trinity Western University, at $16500.
26. York University's course handout kits are classified carbon neutral, which means they are produced using environmentally responsible printing practices that include incorporating locally produced papers manufactured using sound forestry practices and increased recycled fibre content.
27. ***C***
28. Virtually anybody can make biodiesel, because it is so easy you can even make it in your own kitchen.

Document critique—mechanics

(The underlined words indicate where errors have been corrected.)

Let's examine some of the claims made by environmental opportunists such as Carbon Footprint, a company that makes money from selling carbon offsets.

> **Claim number one:**
>
> "At Carbon Footprint, our belief is that climate change will only be addressed if each and every one of us takes responsible steps to **REDUCE** their [sic] CO2 emissions."

I've been looking for evidence that CO_2 emissions 1.), actually cause global warming and that 2.), global warming is necessarily a bad thing. Here's what I've found at Carbon Footprint's website: I see scientific evidence that temperatures have been rising (0.7 degrees C in the past 100 years, world wide) and that glaciers have been melting (24,000 $kms.^2$ of Arctic ice in 30 years). But I don't see any evidence that released CO_2 has had anything to do with it. Yes, the increase in CO_2 emissions has been linked with warming, with evidence of those patterns going back four hundred thousand years but where's the connection? One has to prove **correlation**, not just timing. You could just as easily say that periodic global warming causes greater CO_2 emissions and that interspersed global cooling caused lower CO_2 emissions.

> **Claim Number two:**
>
> "Even though the hole in the ozone layer doesn't contribute to global warming. We should still be concerned about the affects of increased UV light."

Regardless of their claims Carbon Footprint hasn't proved this either. Why should we just take their word for it? The company doesn't provide no evidence at their website that the "hole in the ozone layer doesn't contribute to global warming". None, Zilch.

> **The important thing to remember about Carbon Footprint:**
>
> "From this web site only - you can offset your CO2 emissions while you shop on-line. Every purchase you make on-line via the links on this page will count towards our carbon offset programme - including reforestation and clean energy projects globally."

So, we see what Carbon Footprints all about – money! When its propaganda makes you feel guilty enough, then you buy its services and you shop with its partners, such as eBay, Amazon books, and napster. Or you buy its Ultra bright windup Eco Torch.

Don't be fooled by what the media give you; don't be fooled by outfits like Carbon Footprint, who are in the business of selling guilt about global warming. Let's put our energy into dealing with the impacts of warming instead of succumbing to the arrogant human assumption that we can cause a whole planet to change its climate.

Vocabulary and Style Exercises

Practice session 1—choose vocabulary that suits the reader's technical level

2. To turn the device on, press and hold the **Power/Lock Key**.

   Why?

   This second choice provides just enough instruction for the reader, although an image of the power/lock key would also help the reader understand exactly what to press. The first choice does not tell what to press in order to turn the device on.

   The third choice does not issue an instruction and then it adds unnecessary technical information that clouds the issue.

Practice session 2—use precise, accurate phrasing

1. Last month our outlet had $103,000 more sales than the Montreal outlet.
2. This car has a top speed of 312 km/h.
3. Workers in this office do not know who is in charge.
4. Your crew damaged a smartscreen projector.
5. We had 43 mm of rain last night.

Practice session 3—use an objective tone

1. The group's spokesman demanded overdue salary increases.
2. We do not see that your request for promotion is justified by your performance.
3. This report finds that the contractor has used sub-par materials in building the bridge and is therefore at fault.
4. The security breach for which you were responsible has led to theft of company equipment.
5. I cannot approve your proposal for a 20 percent increase in our marketing budget.

Practice session 4—use simple, familiar language when appropriate

1. We wish you luck.
2. You should smoke fewer cigarettes.
3. A good writer knows correct grammar.
4. I will try to choose the best candidate.

Practice session 5—use jargon sparingly

1. The X-33 printer will cost our firm $3,000.
2. To make more money, prioritize your investment goals.
3. Fifty percent of customers who try this product use it again.
4. We welcome all new managers.
5. We will use intercoms to involve our substitute employees.

Practice session 6—identify empty jargon and pompous phrasing

agree

I am in agreement with your proposed solution to the overtime problem, which seems increasingly

irritates

our overtime scheduling system

vexatious to both management and the union, because the current way of allocating overtime work

is unfair to employees and is not in keeping with overall management policy. However, I am not

it doesn't match

enamored with the projected costs of your proposed approach, which appears burdensome for our

don't like

is too expensive

company. Therefore, it is my considered opinion that a re-working of the overtime pay scale will

I believe

be necessary, particularly because we are in a heavy debt load situation at the moment.

we'll need to renegotiate overtime pay

the company is deep in debt

Practice session 7—identify wordy phrasing

1. Return that spectrometer to the person named Gordon Burns when you have finished using it.
2. Will the ratepayers make payment of the development costs for the proposed riverside trail system?

3. Some, though not all, air toxins come from absolutely natural sources such as burning forest fires and the wearing away of soil erosion.
4. The Nissan Maxima SE sports sedan automobile that I have for sale is silver in colour.
5. There are several aviation-related businesses that lease land and facilities from YVR.

Practice session 8—reduce wordiness

fewer in number	*fewer*	it would appear that	*apparently*
for the reason that	*because*	final conclusion	*finally*
goes under the name of	*called*	lenticular in character	*lens -shaped*
surround on all sides	*surround*	equally as good	*equal*
if conditions are such that	*if*	oval in shape	*oval*
small in size	*small*	disappear from view	*disappear*
in view of the fact that	*therefore, so*	plants exhibited good growth	*grew well*
I personally feel	*I feel*	prior to	*before*
it is often the case that	*often*	the fish in question	*the fish*
repeat again	*repeat*	serves the function of being	*is*
proceed ahead	*go, proceed*	it is possible that the cause is	*possibly*

Practice session 9—use active verbs instead of weak verbs

1. Our disposal procedure conforms with federal standards.
2. Please decide today.
3. We need discuss the problem.
4. Our conclusions agree.
5. This manual instructs end users.

Practice session 10—write concisely

1. Igor needs a vacation.
2. If the system fails, you must sound the alarm.
3. The committee chairperson recommends these actions.
4. We must accept the offer.

5. Please analyze and evaluate our new system.

Practice session 11—write concisely

1. I admire Professor Singh.
2. We made the lowest bid, so we won the contract.
3. We have worked together before.
4. This project report is the most informative.
5. She works hard.

Practice session 12—eliminate needless prefaces

1. The fire has severely damaged the reactor.
2. Jenna had several reasons for leaving the company.
3. We must act immediately.
4. Clayton has reported several safety violations.
5. I am pleased to accept your job offer.
6. Few people can report so precisely on this topic.
7. Several key issues now confront our company.

Practice session 13—eliminate needless prepositions and use direct phrasing

1. Table 5.12 presents the implementation plan.
2. Previous assessments helped identify potential restoration sites, and assigned preliminary work priorities in some cases.
3. Our preliminary research makes us confident that our plan will benefit the environment.
4. Researching this report was useful for learning about project management.
5. The methods described above, with a crew on site during critical draw down periods, will minimize fish mortality.
6. Replacing and modifying key components, and ignoring the display decimal allowed the speedometer to read the wire feed rate in inches per minute.

Practice session 14—eliminate ambiguities

1. Call me after 7 o'clock any evening except Tuesday.
2. I can recommend this candidate very highly. or, I cannot recommend this candidate.
3. Janice dislikes working with Claire because Claire's impatient.
4. Despite Joe's efforts, he misinterpreted Sam's message.
5. Our division's writers needs to be more effective. Or: Our division needs more writers who are effective.

6. Tell the reactor operator to sound a general alarm and then leave the building.

Practice session 15—clarify ambiguous modifiers

1. Determined to maintain his authority, the professor did not accept late assignments.
2. After a hard day's work, many people like to relax in front of the TV.
3. Using logical arguments, the employee persuaded his boss to grant a pay raise.
4. I want your division to write the final Amex report. Or: I want the final Amex report that your division wrote.
5. He ordered only six laptops for the new office.

Practice session 16—identify paragraphing techniques

1. Circle the topic statement in the following paragraph.
   - Does that statement cover the paragraph's content? **Partly**
   - Does it signal the paragraph's purpose? **No**

You'll like our floors. A quality hardwood floor adds ~~warmth, elegance~~, and increases the

You'll like our floors for a variety of reasons.

warmth and elegance

value of your home. It strengthens and stabilizes the inherent structure of your home. ~~Because of~~

For example, hardwood

~~ease of maintenance, and hardness,~~ wood flooring allows you to have the healthiest of living

Also, because its hard surface is easy to maintain,

environments, ~~Also,~~ hypoallergenic. ~~Our~~ company offers a full range of wood species, colours,

and it is

You'll be glad to hear that our

sizes, ~~as well as~~ finishes to meet the needs of the most discerning wood flooring buyer. Each floor

and

we deliver ~~reflects the many years of development in state of the art production, our Premium~~

> has been flawlessly milled and finished with our Premium finish, which provides the best wear characteristics on the market.

~~finish is the most wear resistant finish on the market.~~

Note: The proper spelling of "state of the art" is "state-of-the-art."

Practice session 17—use active voice instead of passive voice

1. RJ Overton and Bliss Strand presented the report to the board on May 5.
2. Our department places high priority on nuisance weeds.
3. The well produces excellent water quality.
4. However, our study of local topography and surface water features, suggests that groundwater moves from east to west.
5. Our department will prepare a final report when we receive your comments.

Practice session 18—use passive voice for special circumstances

1. *Edited to emphasize the recipient:* The completion dates were estimated by our project design team.
2. *Edited to emphasize what was done:* Woody debris, boulders, and gravel was removed from the site by Channel Construction, following a storm in May 2014.
3. *Edited to emphasize what happened rather than who made it happen:* In the 2013 Texas hold 'em Grand Finale, $1.7 million was won on the very last hand with a Royal Flush after Harvey Grace frequently bluffed his opponents in earlier deals.
4. *Edited to emphasize what is being provided:* Recommendations for mitigating potential impacts of the bridge construction are presented.
5. *Edited to emphasize the result:* Improved access to the beach will not be created by the proposed work.

Practice session 19—place the action in verbs, not in nominalized nouns

1. Please describe each of the slope failures.
2. Activate the compression valve to rapidly fill the chamber.
3. Once we verify fish habitat, we can decide how to design the channel to ensure that we minimally disrupt the spawning grounds.
4. We recommend how to mitigate the potential impact of building the bridge.
5. Please respond as soon as possible.

6. She was unable explain the bearing failure.

Improvement exercises—sentences

1. The report would enable the Maintenance Department at CanaWood's Kelowna plywood plant to show upper management that the roller screw is feasible in this application, so that management will approve in-plant testing.
2. We researched the Microsoft's ISA server, a software firewall that is installed on a computer or server.
3. The following progress report answers your request about the status of our study of wastewater management in Oyama, which is being prepared for the Oyama Concerned Citizens Committee.
4. The report will review the community's wastewater management issues and recommend further progress in reducing nutrient loading from wastewater discharges to area lakes.
5. The following summarizes the project's progress to date.
6. The assessment met Natural Resources Canada's requirement for its project financing.
7. When one understands the job requirements and roles in shipping, one can see the advantages of shift work.
8. The following memo describes the activities of the proposed workshop series.
9. Our research about your marketing tools identified a security hole in your external network. This discovery helped us create a software tool that is easy to use, but powerful enough to prevent security breaches.
10. I have often attended your concerts.
11. The writing instructors assess participants' writing skills and use their knowledge, teaching experience, and available computer resources to help those participants who don't communicate well.
12. So the recent drainage control improvements along West Sitkum Road, which have completed most of the 1996 assessment's recommendations, have minimized the risk of sedimentation in local water domestic water supplies.
13. Our website requires an interactive inventory.
14. A few of the plant species were unavailable at the time of planting, so we planted more of the available species.
15. The College computer laboratory facilities offer specialized training.
16. Water quality should be monitored in Richmond Creek, upstream and downstream of the proposed outfall site, beginning two years before the plant starts up.
17. Both species prefer colder water, which usually occurs in this lake.
18. Our mortgage software automatically calculates the effects of rising interest rates, your income, and various amortization periods. The software then coordinates with fluctuating property values and property taxes and compares the results to your pre-approved credit

rating and possible payment schedules. Finally it requests quotes from 18 Canadian lenders.

19. The study area is not close to the Fraser River.

20. An important part of this project will be to establish goals and report whether these goals have been met. This process will help ensure total accountability and transparency.

Improvement exercise—paragraph

Cork has been a flooring material for over one hundred years because of its features:

- it provides outstanding acoustical and thermal insulation;
- it cushions the foot;
- it resists moisture damage and decay; and
- it has proved very durable, as evidenced by manufacturers' warranties that cover all damage except water damage.

Also, cork is easy to clean because of its five coats of UV-cured acrylic finish that contains no volatile organic compounds. A light cleaning with a damp mop will maintain the floor's new appearance.

Document critique 1—structure, phrasing, and mechanics

The Woodsman 993 hybrid set of irons bridges the gap between irons and hybrids, as designed by Ed Ring. In order to create an evenly spaced set that has hybrid characteristics throughout the set, he increased the lofts on all the irons, he replaced the 3 iron and 4 iron with hybrids, and he gave the remaining eight irons hollow heads and extreme perimeter weighting.

The 993's lofts vary from traditional settings. The hybrid 3 and hybrid 4 have 20 degrees of loft and 22 degrees, respectively, which is four degrees stronger than standard. The lofts on the 5, 6, and 7 irons are also four degrees stronger than standard, at 24, 27, and 30°. However, starting with the 8 iron, the set starts to come closer to standard specifications: the 34-degree 8 iron is 3 degrees stronger, the 9 iron is two degrees stronger, at 39 degrees. Then the pitching wedge is 44 degrees (as compared to the standard 48 degrees), the gap wedge is 50 degrees, and the sand wedge is 55 degrees.

The strong lofts would ordinarily make the ball fly lower than normal, but Ring's design compensates for those lofts. The heads have hollow heads, a thin face, and a huge cavity back that is filled with a light (only 5 grams) carbon composite cap. The 35 grams thus saved have been moved to the bottom and back of the head. Thus, the iron has a low, deep center of gravity that launches the ball high. Combining this high launch angle with less loft than normal makes the ball go farther with each club. A person who would hit a ball 135 yards with a standard 8 iron will hit it 145 or 150 yards with the Woodsman 993 7 iron. That distance gain is not penalized because the ball will still come in nice and high with good spin, so it will stick on the green.

An average player will hit the ball 175 yards with the 5 iron and 190 yards with the 4 hybrid, a useful gap between clubs, depending on which shaft the player chooses. The set is available with Rifle steel shafts or with Aldila graphite shafts, which have been designed to match the 993 heads.

Note: GolfSpyder.com explains "loft" at http://www.golfspyder.com/golf-club-loft.html.

Document critique 2—structure and phrasing

SUMMARY

Smith Consulting Ltd. has conducted a geotechnical investigation of subsurface soil and bedrock conditions at the proposed arena location at the northwest corner of 17th Avenue and Corcoran Road. This report interprets the geotechnical data and recommends an approach based on that interpretation.

In addition to observing the surface conditions, the investigative methods determined moisture content and ran Atterberg Limit tests. These tests confirmed inferences that had been based on the observations and tests of surface conditions.

The investigation revealed that the east half of the proposed building site consists of a bedrock knoll while the west is underlain by 0.5 to 4.0 meters of fill materials followed by up to 0.9 meters of loose organic soils. The fills and/or organic soils are underlain by native deposits of stiff clayey silt and clay.

The bedrock surface will require drilling and blasting to the proposed design grade elevations and the native stiff clayey silt and clay will provide a suitable base for the proposed arena facility. As indicated in Section 4.1, the entire building should be founded on a minimum of 1.0 meter thick of granular fill placed and compacted as specified in Section 4.2. A soil weight of 2000 kilograms per cubic meter may be used in the design calculations.

Observations and borehole results indicate that approximately three-quarters of the proposed arena site is underlain by end dumped fills ranging from 0.3 to 1.8 meters thick followed by a thin layer of surficial organic soils. The above surficial soils and fills are further underlain by an interlayer sequence of clay, silty clay, and clayey silt. Also, the proposed design will include an approximate 3.7 meter high berm around three-quarters of the proposed building. This berm could add 50 to 70 millimeters of recompression settlements. In order to minimize potential differential settlements under the proposed structure, the proposed arena site should be preloaded.

An alternate option for locating the arena favors the north portion of the site where the fill thickness is relatively thin and/or nonexistent. The granular fill that presently overlies the proposed arena site has essentially preloaded the site for an estimated 21 years. Therefore, a pre-foundation treatment of this site would require excavating only to remove any further fills and/or surface organic soils beneath the proposed building. That approach would be cheaper than the option investigated in this report. [391 words]

CORRECTION SYMBOLS

Symbol	Meaning
ab	abbreviation
active	Use active voice instead of passive voice
address	Change address (1st person, 2nd person, or 3rd person point of view)
amb	ambiguous phrasing
bias	biased tone
ca	faulty pronoun case
cap	capitalization
coh	paragraph lacks coherence (doesn't flow; lacks transitions)
cs	comma splice
coord	faulty coordination
DM	dangling modifier
emph	change emphasis
frag	sentence fragment (incomplete thought)
jarg	needless jargon
ind ph	indirect phrasing (make it direct!)
len	paragraph length
lev	level of technicality
logic	faulty logic
MM	Misplaced modifier
mng	meaning unclear
om	omitted word
over	overstatement
par (or //)	faulty (or lack of) parallelism
p	punctuation error
P (or ¶)	start new paragraph
pomp	pompous phrasing
prep	wrong preposition or excess prepositions
pro agr	Pronoun doesn't agree with antecedent

pro ref (or ref)	faulty or vague pronoun reference
ro	run-on sentence
seq	sequence of development in a paragraph
sexist	sexist usage
sp	spelling error
sub	faulty subordination
s/v agr	subject/verb agreement
th op	"there is" or "there are" sentence openers
trans	transition—faulty or needed
ts	topic sentence
un	paragraph lacks unity
var	more sentence variety required
w	wordy phrasing
wo	word order problem
ww	wrong word
#	problem with numbers usage

APPENDIX A: Punctuating Lists

The way we should punctuate a list depends on the nature of that list. For example, **a simple list** of single items or short phrases may not require internal or end punctuation.

The College has six Business Instructors at its Downtown Campus:

- Joyce Gruber
- Rocky St. Pierre
- Magnus Halvorson
- Rita O'Toole
- Davinder Bains
- Yee Ling Chan

However, the writer who wishes to present **the list as a complete sentence** that's easier to read than a normally formatted sentence would need to include the following punctuation and coordinating conjunction.

The College has six Business Instructors at its Downtown Campus:

- Joyce Gruber,
- Rocky St. Pierre,
- Magnus Halvorson,
- Rita O'Toole,
- Davinder Bains, and
- Yee Ling Chan.

Let's say that the **list contains phrases of more than two words**. One choice would use a series of commas, a conjunction, and end punctuation.

The College now includes several diploma programs:

- a comprehensive Business Administration program,
- three trades programs,
- an aviation maintenance course, and
- a new nursing program.

But—there seems to be an exception to every rule! The following list, which appears in the 6th Canadian edition of *Technical Communication*, doesn't punctuate each phrase. The authors envisioned **a simple list, not an integrated sentence**.

Although you might not anticipate a career as a "writer," your writing skills will be tested routinely in situations like these:

- proposing various projects to management or to clients
- contributing articles to employee newsletters
- describing a product to employees or customers
- writing procedures and instructions for employees or customers
- justifying to management a request for funding or personnel

- editing and reviewing documents written by colleagues
- designing material that will be read on a computer screen

Another example (that **requires additional internal punctuation**) could use a series of semicolons that would correspond to the way semicolons would be used in a complex sentence. First, here's that "normal" sentence:

> The College now includes several diploma programs: a comprehensive Business Administration program, offered annually; three trades programs, offered as demand requires; an aviation maintenance course, at the airport; and a new nursing program, slated for 2009.

That sentence would be easier to read in a bulleted list.

> The College now includes several diploma programs:
>
> - a comprehensive Business Administration program, offered annually;
> - three trades programs, offered as demand requires;
> - an aviation maintenance course, at the airport; and
> - a new nursing program, slated for 2009.

Lists of clauses can use semicolons, as in the following **integrated sentence list**.

> The College now includes several diploma programs:
>
> - a comprehensive Business Administration program is offered annually;
> - three trades programs will be offered as demand requires;
> - an aviation maintenance course has been started at the airport; and
> - a new nursing program is slated for 2009.

Or normal end punctuation might be used, as in this example from *Technical Communication*, 6th Canadian edition.

> To a large extent, that information is stored and transmitted electronically. Several electronic technologies, collectively known as information technology (IT), enhance the speed, volume, and variety of means of creating and transmitting messages:
>
> - Integrated software facilitates the inclusion of verbal, graphic, and video elements from word processors, internet sources, digital recording and storage media, electronic spreadsheets and databases, and oral presentation software.
> - Basic email helps people exchange ideas and information, while email attachments allow for sophisticated formatting of documents.
> - Wireless laptop computers allow users to receive and send email from locations (hotel conference rooms, airport lounges, classrooms, etc.) that support wireless transmission.

Bulleted or numbered lists of sub-paragraphs present few punctuation challenges—each sub-paragraph uses its "normal" punctuation. The following list appears in the 6th Canadian edition of *Technical Communication.*

> These technologies have had a profound effect on office communication:
>
> - Small and medium-sized companies, which comprise 95 percent of all Canadian businesses, are slower than large companies to adopt emerging technologies such as VoIP, wireless networking, and videoconferencing. However, almost all have desktop computers and e-mail access, and many have laptops, PDAs, cell phones, and websites. (Lima "Small firms ..." E2)

- Instead of being housed in one location, the virtual company may have branches in widespread locations, or just one central office, to which employees "commute" electronically. Such arrangements require workers to be skillful communicators who must master the latest technologies, use proper e-communication etiquette, and know how to compensate for the lack of face-to-face contact (Marron C1). A special class of worker has emerged—the "virtual assistant," who works on contract, at home, for one or more firms. A virtual assistant might answer phones, direct e-mails, produce and edit documents, update websites, or maintain databases and inventories. He/she can even go into a client firm's computer (via an internet connection) to work on material stored on that computer's hard drive. (Buckler E5)
- Instead of relying on secretaries, most managers compose and send their own messages via email. Many also compose, design, and deliver their own reports and proposals.
- On desktop publishing networks, the composition, layout, graphics design, and printing of external documents and Web pages are done in-house.
- Paper documents (such as résumés) can be optically scanned and stored electronically. Many other documents are stored as PDF files.
- Computer-supported cooperative work systems, instant messaging, and videoconferencing enable employees worldwide to converse in real time. Email listservs announce daily developments in prices, policies, and procedures.

APPENDIX B: Accidental Humour

The following sentences weren't written by gag writers employed by Rick Mercer or David Letterman. These examples of accidental humor have appeared in student assignments and in actual workplace documents. In the end, though, each bit of levity isn't all that funny because it reflects badly on the person responsible for the mangled English.

- *The trends set by music itself and the natural transitions it goes through is also strong evidence that popular music is popular today.* (This hilarious bit of circular "reasoning" reminds me of the CBC half-time interview of a Winnipeg Blue Bomber defensive back who proudly said during the interview: "Football! I love football. If it wasn't for football, I wouldn't be in football today!")
- *In light of my accomplishments and skills I demonstrated; Wal-Mart promoted me to the curtsy desk. Working the curtsy desk is a real pleasure because it is so demeaning.* (This amazing "confession" appeared in an actual application letter for a marketing position. I think the writer meant "courtesy desk" and "demanding," not "curtsy desk" and "demeaning." At least, I hope so. By the way, the writer's application was unsuccessful.)
- *Never place a power chord within a mouse.* (This sensible admonition appeared in an electronics student's project manual. When I first read the above sentence, I imagined some 120 dB Metallica power chords bouncing a computer mouse around.)
- *I am 17 years old and I have been on a motorcycle since I was 5 years old.* (Life hasn't been easy for this writer! For example, just imagine his saddle sores.)
- *There are about 60 passengers aboard on each cruise. The single tourists make up 66 percent, of which a little more than half are usually male.* (But apparently, they're not <u>always</u> male!)
- *This industry reached a new plato in 2007.* (While writing about adventure tourism, this writer became rather philosophical.)
- *The job market is really tough these days, especially for those in child care, so we have to bare with it.* (This sentence in a five-page job search plan made me wonder whether the writer was paying for her child care studies by working as a stripper.)
- *I also enjoy a strong bake ground in english and have a firm grasp for writing.* (This job applicant may have a strong grip, but he can't spell.)
- *Our trip to southern California and Mexico showed us the full spectre of the hospitality industry.* (What are the odds that this person changed careers after his hotel management class's extended field trip?)
- *Will detonation skiers visit Intrawest's new resort?* (We certainly hope not!)